Cytogenetics: Techniques and Applications

Cytogenetics: Techniques and Applications

Edited by **Morgan Key**

New York

Published by Callisto Reference,
106 Park Avenue, Suite 200,
New York, NY 10016, USA
www.callistoreference.com

Cytogenetics: Techniques and Applications
Edited by Morgan Key

International Standard Book Number: 978-1-63239-144-5 (Hardback)

Printed in the United States of America.

Contents

Preface

This book discusses the current trends in the field of cytogenetics. It involves the specifics and particulars of the methodologies that can be accepted and used in clinical laboratories. The book discusses the basic methods of cell lines, primary cultures and their usage; array CGH for diagnosis of fetal conditions; micro technologies and automations; use of chromosomes as tools to find biodiversity and usage of digital image technology. It also includes methods to deal with acute lymphoblastic and myeloblastic leukemia in patients and survivors of atomic bomb exposure. While focusing on the advanced practices used in this field and its applications, this book demonstrates the urgency to establish cytogenetic labs with modern and latest equipment. This book will aid in accurate and effective diagnosis which will further benefit patients.

Various studies have approached the subject by analyzing it with a single perspective, but the present book provides diverse methodologies and techniques to address this field. This book contains theories and applications needed for understanding the subject from different perspectives. The aim is to keep the readers informed about the progresses in the field; therefore, the contributions were carefully examined to compile novel researches by specialists from across the globe.

Indeed, the job of the editor is the most crucial and challenging in compiling all chapters into a single book. In the end, I would extend my sincere thanks to the chapter authors for their profound work. I am also thankful for the support provided by my family and colleagues during the compilation of this book.

Editor

Cytogenetic Analysis of Primary Cultures and Cell Lines: Generalities, Applications and Protocols

Sandra Milena Rondón Lagos[1] and Nelson Enrique Rangel Jiménez[2]
[1]*Doctoral Program in Biomedical Sciences, Universidad Del Rosario*
[2]*Azienda Ospedaliero-Universitaria S. Giovani Battista di Torino*
[1]*Colombia*
[2]*Italy*

1. Introduction

Cytogenetics constitutes an important diagnostic tool to determine and/or confirm specific syndromes nowadays; its use is directed towards the selection of treatments and monitoring of patients using different procedures. These latter are carried out in order to obtain a karyotype from peripheral blood or several tissue biopsies (e.g. biopsies from patients with melanoma, breast cancer, skin biopsies, foreskin samples, abortion products, among others). However, the study of chromosomal abnormalities in culture cells has been limited by complex processes such as achieving cell growth and a good number of metaphases, which in turn hampers the chance to obtain a useful number of metaphase spreads in order to carry out a proper cytogenetic analysis, that should be able to display a good morphology, an adequate dispersion and a correct banding. Cell lines are widely used in different research fields, particularly in invitro models for cancer research. (Burdall et al., 2003)

Given the importance of the model used to examine and manipulate potentially relevant molecular and cellular processes underlying malignant diseases, it is necessary to achieve an accurate and comprehensive karyotyping for cultures of different cell lines. In turn, karyotyping provides an insight into the molecular mechanisms leading to cellular transformation and could allow clarifying possible cytogenetic aberrations associated to drugs exposure and the development and progression of different types of cancer. The fast increase observed in cancer incidence is forcing us to carry out more identification studies of cytogenetic biomarkers associated to development of this disease, which could contribute for a better understanding of the carcinogenic process and could also have enormous implications for the development of effective anticancer therapies.

Obtaining metaphase cells for chromosome analysis requires the use of a series of reagents, protocols, and environmental conditions, among others, that will allow us to collect the chromosomes. Metaphase cells must be cultured under certain conditions in order to obtain a proper number of dividing cells, which need to grow and divide fast in this medium as well. Taking into account all the issues mentioned above, an accurate knowledge from the culture medium and its conditions, as well as the techniques and protocols required in general, will allow a good cell growth in the culture and the collection of chromosome spreads to carry out the cytogenetic characterization and the identification of chromosomal abnormalities present in the cells in study.

This chapter describes the practical aspects of performing cytogenetic studies in primary cultures and human cancer cell lines that have been previously standardized, in order to be applied not only in research but in diagnosis and possible treatment of several diseases.

2. Generalities

The application of cytogenetic studies on tissue and cancer cell lines has become important in recent years, because the presence of some chromosomal abnormalities indicate the prognosis of the disease and the corresponding response to therapy. The most common clinical applications of cytogenetic studies on tissue and cancer cell lines are:

- Establishing the type of chromosomal abnormalities and its frequency
- Identifying the genes located in the affected chromosomal regions in order to establish those possibly implied in neoplasia
- Studying tumorigenic and metastatic behaviors, apoptosis and functionality
- Identifying the mechanisms of action used by hormones
- Establishing models for drug resistance studies
- Establishing the therapeutic potential of different treatments
- Supporting further research.

The knowledge of novel chromosome rearrangements and breakpoints identified could be useful for further molecular, genetic and epigenetic studies on human cancer that could lead us to understand the mechanisms involved in the development and progression of this disease.

2.1 Characteristics of cell cultures

Cell cultures can be divided into two groups, depending on the substrate used for cell growth:

- Suspension cultures: Cells are cultured by constant agitation in a liquid medium. Cell cultures are prepared by diluting cell suspensions.
- Monolayer cultures: Cells adhere to a solid (glass or plastic) or semisolid (agar, blood clot) surface, forming a cell surface, which can be observed by light microscopy or phase contrast. Cultures are maintained by releasing cells from the substrate using mechanical or enzymatic procedures, continuing their life cycle in new cell subcultures.

2.1.1 Requirement of cell growth

There are several variables that determine whether a cell will multiply *in vitro* or not, some of these depend directly on the conditions of the growth medium and some do not.

- The growth medium must possess all the essential, quantitatively balanced nutrients. It must include all the necessary raw material to promote the synthesis of cellular macromolecules; it must also provide the substrate for metabolism (energy), vitamins and trace minerals (their primary function is catabolic) and a number of inorganic ions implied in the metabolic function.

- Physiological parameters: temperature, pH, osmolarity, redox potential, which must be kept within acceptable limits.
- Cell density and subculture mode
- Serum is added to the basal medium to stimulate cell multiplication and interaction with the other variables contained in the system. It serves as a source of macromolecular growth factors. Serum is a very effective supplement that promotes cell division because it contains different growth-promoting factors. Complete serum contains most of low molecular weight nutrients required for cell proliferation. Serum may neutralize trypsin and other proteases, provide a protein "carrier" to solubilize water-insoluble substances (such as lipids) and has the ability to provide hormones and growth factors to cells.

2.1.2 Contamination of cell cultures

Cell cultures can be contaminated by fungi, bacteria, mycoplasma, viruses, parasites or cells from other tissues. It is mistakenly thought that tissues obtained using aseptic techniques from apparently healthy animals are sterile; however, it is common to find bacteria, mycoplasma, viruses or other microorganisms in these tissues. Fungi and bacteria are universally distributed in nature and are relatively resistant to environmental factors such as temperature, radiation, and desiccation, among others.

These organisms can appear in cultures due to several factors:

- Through dust particles carried by air currents.
- Aerosols produced by the operator during handling.
- Through non-sterile equipment.

Viruses and mycoplasmas are found in nature mainly in cells and body fluids, and these are more sensitive than fungi and bacteria to environmental factors. The most important sources of contamination with mycoplasma are aerosols and sera used in culture mediums. Other routes of entry for the virus are other infected cell cultures, serum or spray. Three factors are determining the effectiveness of a sterility test:

- Sensitivity and spectrum of the medium used.
- Incubation terms and time.
- Sample Size

The medium used for these tests should be sensitive and have a broad spectrum to detect anaerobic bacteria, fungi and mycoplasmas in routine testing. Cultures for bacteria and mycoplasma should be incubated aerobically and anaerobically, in order to avoid the loss of detection of some microorganisms. It is recommended to test for sterility at different times in the initiation and harvesting of the cell culture (beginning, middle and end).

2.1.3 Contamination of cell cultures by other cells

A very common contamination, generally not considered by researchers working in cell culture, is cross-contamination between cell cultures, both at the intra and interspecific level (MacLeod et al., 1999; Marcovic & Marcovic 1998; Masters et al., 2001; van Bokhoven et al., 2001; Masters 2002). Several cases of cross-contamination between cell cultures have been documented in the last years, this has been possible by some sources that are able to provide certified cell lines, which can be used when contamination in the cell culture is suspected. Approximately, a 20 to 30% of cell cultures are contaminated at the intra or interspecific level, and it is believed that this value is higher due to the large number of contaminated cultures, on which there is no suspicion. The most convenient way to avoid contamination is to use rigid sterilization and aseptic techniques, the culture medium must be proven for contamination before use, working with cell cultures in laminar flow chambers, decontaminate the work area on a daily basis and furthermore, when there is manipulation of different cell lines.

2.2 Human cancer cell lines

Many human carcinoma cell lines have been developed and are widely used for laboratory research, mainly in studies of tumorigenic and metastatic behaviors, apoptosis, functionality, and therapeutic potential, and particularly as in vitro models for cancer research. Among these cell lines are the following: MCF-7, SKBR3, TD47 and BT474.

2.2.1 Characteristics

MCF-7 is a cultured cell line from human breast cancer, which is widely used for studies on breast cancer biology and hormones' mechanism of action research. The cell line was originally derived at the Michigan Cancer Foundation from a malignant pleural effusion found in a postmenopausal woman with metastatic breast cancer. The cells express receptors as biological responses to a variety of hormones including estrogen, androgen, progesterone, glucocorticoids, insulin, epidermal growth factor, insulin-like growth factor, prolactin, and thyroid hormone with non-amplified HER2 status (Osborne et al., 1987)

The cell line SKBR3 is a highly rearranged, near triploid cell line, derived by Fogh and Trempe (1975) from a pleural effusion and overexpresses the HER2/c-erb-2 gene product. This cell line shows only a weak ESR2 (ERß) expression and no ESR1 (absence of functional ERα) and PGR expression, indicating that this cell line represents models of estrogen- and progesterone-independent cancers, with capability for local E2 formation and possible action via non-ER mediated pathways. ERß expression level in tumor cell lines is characterized by a significantly slowed proliferation (Hevir et al., 2011). ERß may negatively regulate cellular proliferation, promote apoptosis and thus may have not only a protective role in hormone-dependent tissues, such as breast and prostate, but also a tumor-suppressor function in hormone-dependent tissues (Lattrich et al., 2008).

Human breast ductal carcinoma BT474 cell line was isolated by Lasfargues et al (1978). It was obtained from a solid, invasive ductal breast carcinoma from a 60-year-old woman; cells were reported as tumorigenic in athymic mice and were found to be susceptible for mouse mammary tumor virus, confirmed as human with IEF from AST, GPDH, LDH and NP (Lasfargues et al., 1979).

T47D is a cell line derived from human ductal breast epithelial tumor, it was isolated from a pleural effusion obtained from a 54 year old female patient with an infiltrating ductal breast carcinoma (Keydar et al., 1979). These cells contain receptors for a variety of steroids and calcitonin. They express mutant tumor suppressor protein p53 protein. Under normal culturing conditions, these cells express progesterone receptor constitutively and are responsive to estrogen. They are able to lose the estrogen receptor (ER) during long-term estrogen deprivation *in vitro*. Culture conditions, receptor status, patient age and source and tumor type for each cell line are shown in Table 1 and Figure 1.

Cell line	Source code	Passage no	Receptor status	HER-2 status	Tissue source	Tumor type	Patient age	Culture conditions
T47D	ATCC: HTB-133	P20	ER+ PR+	Negative	PE	IDC	54	RPMI 1640 + 10% FBS + 2 mM L-glutamine +antibiotic-antimycotic solution (1X)
MCF-7	ATCC: HTB-22	P16	ER+ PR+	Negative	PE	AC	69	
SKBR3	ATTC: HTB-30	P15	ER− PR−	Positive	PE	AC	43	
BT474	ATTC: HTB-20	P12	ER+ PR+	Positive	IDC	IDC	60	DMEM + 10% FBS + 2 mM glutamine + antibiotic-antimycotic solution (1X)

Table 1. Characteristics of the breast cancer cell lines AC, adenocarcinoma; IDC, invasive ductal carcinoma; PE, pleural effusion; P, passage number. Media conditions: FBS, fetal bovine serum; DMEM, Dulbecco's Modified Eagle's Medium. Cell lines were maintained at 37°C and 5% CO2 in the indicated media

2.2.2 Cytogenetic abnormalities found in human cancer cell lines

MCF-7 cell line has a modal number from 82 to 86 with 56 types of aberrations: 28 numerical and 28 structural aberrations. The most common aberrations in MCF-7 cells are der (19) t (12;19)(q13;q13.3) and add(19)(p13) (Figure 2A).

SKBR3 cell line has a modal number from 71 to 83, with 48 types of rearrangements: 27 numerical and 21 structural rearrangements. The most common aberrations in this cell line are del(1)(1p13) and add(17)(17q25) (Figure 2B).

BT474 cell line demonstrated to have a modal number from 65 to 106, with 67 different rearrangements: 35 numerical and 32 structural aberrations. The most common aberrations in this cell line are: Additional material of unknown origin on chromosome 14: add(14)(q31), derivatives from chromosomes 6: der(6)t(6;7)(q25;q31) and 11: der(11)t(8;11;?)(q21.1;p15;?), losses in chromosomes 15, 22 and X chromosome and a gain on chromosome 7 (Figure 2C).

T47D cell line have a modal number of 57 to 66, with 52 types of rearrangements: 26 numerical and 26 structural. The most common aberrations in this cell line are: der(X)t(6;X)(q12;p11);   der(8;14)(q10;q10);   del(10)(p11.2);   der(16)t(1;16)(q12;q12) dup(1)(q21q43) and der(20)t(10;20(q21;q13.3) Figure 2D. The cell lines SKBR3 and BT474 exhibited amplification of HER-2 gen by FISH and the cell lines MCF-7 and T47D not have amplification for this gene.

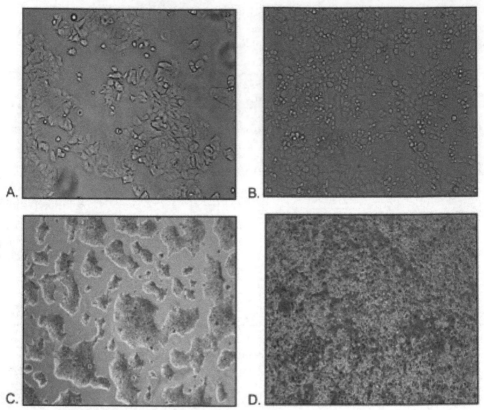

Fig. 1. Inverted microscopic pictures of representative breast cancer cell lines in a monolayer culture. A) MCF-7; B) SKBR3; C) BT474; D) T47D

3. Cytogenetic techniques from tumoral tissue samples and cancer cell lines

Obtaining metaphase cells for chromosome analysis requires the use of a series of reagents that will allow us to collect the chromosomes. Metaphase cells must be grown *in vitro* under certain conditions in order to obtain a proper number of dividing cells. Cells used for chromosome collection must be able to grow and divide fast in the culture medium. Different types of cells may require specific growth factors and medium supplements; once the basic requirements for each cell type are known, the appropriate culture medium is selected, checking sterility appropriately. After the culture has reached the 80% of confluence, it must be harvested and fixed to make a cytogenetic suspension. Cultures are growth arrested and accumulated in metaphase or prometaphase by inhibiting tubulin polymerization and thus preventing the formation of the mitotic spindle (e.g., using colcemid or velbe). Following exposure to colcemid or velbe, cells are treated with a hypotonic solution to enhance the dispersion of chromosomes and fixed with carnoy fixative (Methanol: Acetic Acid). Once fixed, the cytogenetic preparation can be stored in cell pellets, under fixative conditions and 20°C for several months. Fixed cells are spread on slides and air-dried, to be finally banded for the correct identification of chromosomes. Obtaining an

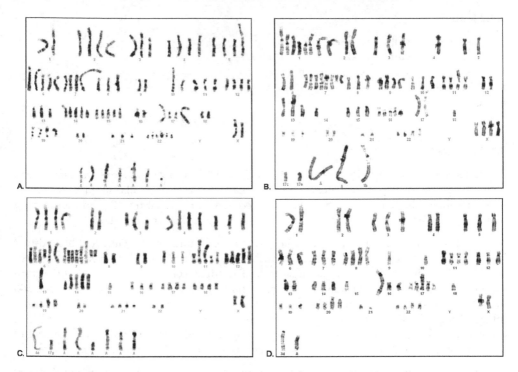

Fig. 2. Karyotypes from breast cancer cell lines. A) MCF-7; B) SKBR3; C) BT474; D) T47D15-ml conical centrifuge tube

adequate quality on chromosome spreads is multifactorial; this will be discussed in detail further on. The amount of metaphases obtained is sometimes inadequate for chromosome analysis, thus it is always necessary to keep growing the cell line.

3.1 Materials, reagents and equipment

3.1.1 Equipment

- Laminar flow chamber
- Incubator
- CO2 Incubator
- Serological bath
- Centrifuge
- Refrigerador
- Microscope with camera
- Inverted microscope
- Magnetic Stirrer
- Micropipettes
- Analytical balance
- Vacuum Pump

3.1.2 Materials

- 75-cm^2 tissue culture flasks
- 25-cm^2 tissue culture flasks
- Sterile disposable plastic transfer pipettes
- Glass slides
- Coverslip sheets
- Petri dishes

3.1.3 Reagents

Solutions should be kept in the dark at -20°C or 4°C, according to the manufacturer's instructions. According to frequency of use, reagents should be aliquoted and frozen. Reagents should be thawed before use and stored at 4°C. Frequent freezing and unfreezing may cause alteration of the culture medium by inactivating the components.

Medium: The most commonly used medium for cell cultures are Dulbecco's modified Eagle's medium-DMEM, RPMI 1640 and DMEM-F12, among others. If the medium does not contain Glutamine, L-glutamine should be added (final concentration 2mM); this is an essential amino acid that is unstable and has a short life at room temperature. To each 500 ml bottle of medium, add 50 ml of Fetal Bovine Serum, 5 ml of L-Glutamine (200 mM) and 5 ml of antibiotic-antimycotic solution (100x). Store the medium up for a month at 4°C. In order to establish primary cultures it is recommended to add also hydrocortisone, estradiol and insulin to the culture medium, providing enough nutrients to induce cell growth.

Serum: Fetal bovine serum; the proportion commonly added is 50ml of serum per each 450 ml of medium. Usually, the presentation of fetal bovine serum is 500 ml, so this amount should be aliquoted in 50 ml aliquots which must be stored at -20°C and thawed at 4°C or room temperature prior to use. It is advisable not to thaw the medium at high temperatures (37°C or more), as this could alter its composition.

Collagenase stock solution: Type 2 collagenase. To make the stock solution, dissolve 215 U/mg collagenase in distilled water to obtain a final concentration of 2000 U/ml, filter the solution through a 0.2-μm filter and prepare 1 ml aliquots, these can be kept stored for 2–3 months at –20°C. The working solution of 200 U/ml is prepared immediately before use, adding 1 ml collagenase each 9 ml of complete medium. This solution should be kept at 4°C.

Arresting agents:

Colcemid: Colchicine inhibits microtubule assembly by binding to a high affinity site on β-tubulin. Colchicine binding occurs in a nearly irreversible manner and exerts a conformational change in tubulin, as well as in colchicine itself. (Daly, et al. 2009). Colcemid is used on cell lines displaying a high-speed replication and is applied to a final concentration of 0,01 μg/ml for 2.5 hours.

Velbe: Described as a vinca alkaloid, also called vinblastine, this agent is derived from the periwinkle plant, *Catharanthus roseus*, and is noted as the most successful anticancer agent within the past few years. Binding of the vinca alkaloids to β-tubulin occurs fast and reversibly at an intermolecular contact point (Daly, 2009). It is recommended to use Velbe if the rate of cell replication is low at a final concentration of 0,01 μg/ml in a maximum of 16 hours.

The application of these reagents can arrest cells in metaphase and helps chromosomes contraction, allowing an easy recognition of these cells in pro-metaphase or metaphase. The use, exposure time and Colcemid or Velbe concentration varies and depends on several factors, including cell type and overall growth characteristics.

Hypotonic solution: Saline solution that allows chromosomes dispersion within the cell membrane, facilitating its observation and recognition. In order to obtain chromosome preparations from cell lines, the following hypotonic solutions can be used; the selection of this solution will depend on the degree of chromosome condensation obtained.

0.075 M potassium chloride (KCl): Use 5.59 g KCl and make up to 1 liter of aqueous solution. Use the solution at 37°C.

20 mM potassium chloride (KCl) and 10 mM sodium citrate ($Na_3C_6H_5O_7$): Use 1 g KCl and 1g sodium citrate and make up to 500 ml of aqueous solution. Use the solution at 37°C. Its use is recommended with longer chromosomes, that may be twisting or overlapping.

Fixative: Reagent used to stop the action of hypotonic solutions and which in turn, has several functions throughout the procedure related to hemolysis, dehydration, chromosomes fixation and removal of debris membrane that may interfere with the chromosome extended. This reagent is prepared with three parts of absolute methanol and one part of glacial acetic acid. This should be freshly prepared just before its use and should be kept always cold (-20°C).

10x Trypsin-EDTA: Stored frozen in 1 ml aliquots. Diluted 1:10 in PBS when required to obtain a 1x working solution. Store indefinitely at 4°C. Place at room temperature or 37°C before use.

Phosphate-buffered saline (PBS): pH 7, used for diluting solutions.

Stains:

Wright's stain: This stain is usually obtained as a powder. Cover a flask with aluminum foil and insert a magnetic stirrer. Add 0.5 g stain and 200 ml methanol. Stir for 30 min. Filter using a filter paper into a foil-coated bottle. Close the lid tightly and store the bottle in a dark cupboard for at least a week before its use. The stain should be diluted immediately before use at 1:4 with pH 6.8 buffer.

Giemsa: This stain is usually obtained as a liquid. Before use, the following mixture must be prepared: 0.2 ml Giemsa, 0.2 ml Sorensen Tampon and 4.6 ml water (the amount used to dye a slide).

Saline-sodium citrate (SSC) buffer: This is a widely used weak buffer, which is used to carry out several washes and to control stringency during *in situ* hybridization. The 20x stock solution consists in mixing 3M sodium chloride and 300mM trisodium citrate. To make the stock, dissolve 38,825 g sodium chloride (NaCl) and 22.05 g sodium citrate ($Na_3C_6H_5O_7.2H_2O$) in 200 ml of water. Adjust to pH 7 with NaOH or HCl if necessary, make up to 250 ml and sterilize by autoclaving procedures.

Sorensen Buffer: This buffer is used for G-Banding. The working solution consists in two solutions: KH_2PO_4 and Na_2HPO_4. Prepare the buffer as follows:

- Sln A: KH_2PO_4. Dissolve 4559 grs in 500 ml of sterile distilled water

- Sln B: Na_2HPO_4. Dissolve 4755 grs in 500 ml of sterile distilled water.
- Take 500 ml of solution A and mix it with 496.8 ml of solution B, keep it at 4°C.

HCl 0,2 N: Used for G-Banding. To prepare 1000 ml of solution, add 8,25 ml HCl 37% and 500 ml H_2O into a glass container. Store at room temperature.

3.2 Cell culture methods from tumoral tissue

To ensure cell growth and obtain cells in metaphase, it is important to take into account all the sampling conditions. Sterile, non-necrotic tumor samples must be collected in a transport container using optimal conditions of sterility; for example, a sterile tube containing sterile culture medium, an antimycotic and a double concentration of antibiotics, which should be transported to laboratory facilities under controlled temperature.

The tissue sample must be representative, sterile, and viable. To ensure fast cellular growth and prevent contamination with other cell types, the cultures must be incubated in a small culture flask ($25cm^2$) or directly on microscopic slides mounted in multi-well chambers. Cell attachment, proliferation, and mitotic rate should be monitored by daily x examination through an inverted microscope.

The steps to obtain metaphases are:

3.2.1   Dissociation of solid specimen: Enzymatic and mechanical procedures
3.2.2   Culture initiation
3.2.3   Culture harvesting and metaphases
3.2.4   Banding techniques
3.2.5   Freezing of viable cells

The way of determining the time of harvest, colchicine use and exposure to hypotonic solution will depend on the cell type and its growth rate.

3.2.1 Dissociation of solid specimen

Materials

- Collagenase (2000 U/ml)
- Appropriate culture medium (RPMI 1640, DMEM-F12) containing 10% fetal bovine serum (FBS), antibiotic-antimycotic solution (1X), L-glutamine (2 mM), Hydrocortisone, 17β-estradiol and insulin.
- PBS (1X)
- 25-cm^2 tissue culture flasks
- 15-ml conical centrifuge tube
- 5-ml and 10-ml plastic pipettes
- Petri dishes
- Microscopic slides mounted in multi-well chambers x 6 wells
- Tissue dissection equipment: tweezers, scissors

Procedure

Mechanical Disaggregation (Figure 3)

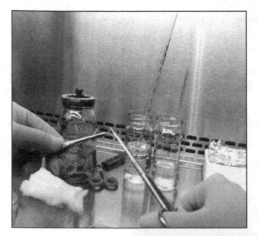

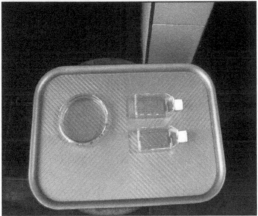

Fig. 3. Representative pictures of tissue cultures, indicating the steps for mechanical dissociation of tissue

- Remove the specimen from the transport container and place it immediately into petri dishes, using 5 ml PBS containing antibiotic and antimycotic agents, in order to wash the tissue.
- Take the tissue and transfer it to another petri dish containing 2 ml of medium; afterwards, remove fat, necrotic tissue and/or blood that may interfere with cell growth.
- Using scissors and tweezers, cut the tissue into fragments of 1–2 mm in size.
- Transfer some of the fragments to a petri dish for enzymatic digestion.
- Take the other fragments and distribute them into 25 cm^2 plastic flasks containing 3 ml of culture medium or in microscopic slides mounted in multiwell chambers x 6 wells containing 1 ml of medium, using glass Pasteur pipettes.
- Place flasks on their sides in an incubator at 37°C in 5% CO_2 incubate for 2-3 days.

Enzymatic Digestion

- Using the tissue fragments previously deposited in petri dishes for enzymatic digestion, add 2-3 ml medium containing collagenase at a final concentration of 200 U/ml.
- Incubate at 37°C in 5% CO2 for 16–24h (overnight), stirring occasionally.

The times used for enzymatic treatment depends on the type of tumor, but generally an overnight incubation is enough. Check disaggregation process under the inverted microscope. A large number of single cells and small clusters of cells may be observed floating at the end of this period.

- After this time, inactivate the enzyme by adding 2 ml of fetal bovine serum (FBS), applied directly to the sample, and transfer the cell suspension to a 15-ml conical centrifuge tube
- Centrifuge for 10 min at 1000 rpm, discard the supernatant.
- Add fresh medium to the tube, mix the suspension by pipetting up and down and transfer the cell suspension to a 25-cm² tissue culture flask or to a multiwell chambers x 6 wells
- Incubate at 37°C in 5% CO2 to allow cells that were attached to the plastic base to grow during the collagenase treatment.

3.2.2 Culture initiation

Materials

- Appropriate culture medium (RPMI 1640, DMEM-F12) containing 10% fetal bovine serum (FBS), antibiotic-antimycotic solution (1X), L-glutamine (2 mM), Hydrocortisone, 17β-estradiol and insulin
- PBS (1X)
- 5-ml and 10-ml plastic pipettes

Procedure

- After an incubation period of 24 hours, monitor the cell cultures (both those who were and were not disaggregated enzymatically). Examine the culture using phase-contrast microscopy to assess the extent of tissue adhesion and cell growth.
- In a laminar flow chamber and under strict conditions of asepsis and sterility, remove the medium containing unattached cells and cellular debris from flasks using a glass Pasteur pipette
- Add gently 2 ml PBS to wash and remove fragments attached; afterwards, remove PBS.
- Add 4-5 ml of the medium and incubate again.

Examine flasks and the multiwell chambers daily through an inverted microscope in order to establish cell growth and mitotic activity. Once cell cultures reach the 80% of confluence, these can be processed to obtain metaphases. Figure 4.

Note: The culture of both fragments (both those dispersed enzymatically and the cell suspension after the enzyme digestion) will ensure cell growth, since in some cases cell growth obtained from cell suspension is insufficient to obtain metaphases

3.2.3 Harvesting of culture and metaphases for chamber slides

In a chamber slide the cells are not removed from the growing surface, is important to control cell confluence, this can not be greater than 80% before the addition of colchicine,

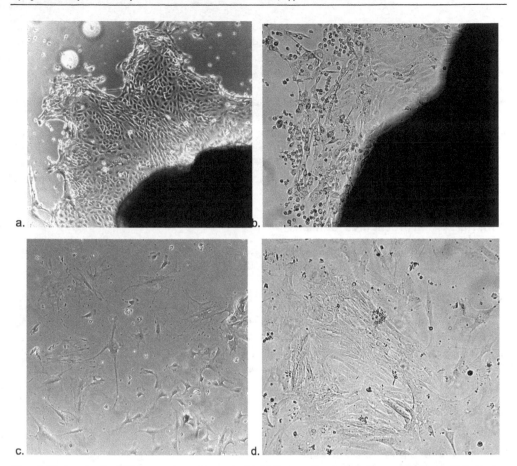

Fig. 4. Pictures of solid tumor in culture. (a, b) Cell growth observed around fragments that were disaggregated only by mechanical procedures, after 6 days of culture (4x-10x). (c,d) Cell growth obtained after enzymatic disaggregation, after 8 days of culture (4x-10x). A good cell growth was observed in both cell cultures; hence, it is advisable to prepare cell cultures using both methods: enzymatically disaggregated fragments and fragments that were mechanically disaggregated only.

since greater confluence is difficult to obtain a good number of metaphases. Colcemid is added to the culture at a final concentration of 0.01 µg/ml for 3 hours.

Materials

- 20 mM potassium chloride (KCl) and 10 mM sodium citrate($Na_3C_6H_5O_7$)
- Fixative methanol–acetic acid (3:1)
- Pasteur pipette

Procedure

- Carefully remove the medium with a Pasteur pipet.

- Add 2ml of prewarmed hypotonic solution (20 mM potassium chloride (KCl) and 10 mM sodium citrate($Na_3C_6H_5O_7$)) slowly down the side of the chamber and put into incubator at 37°C for 17 min.
- Remove the cultures of the incubator and carefully add, around of the well, 1ml of *cold* freshly prepared fixative methanol–acetic acid (3:1) for 10 minutes
- Remove all fluid and add 2 ml of fresh cold fixative slowly down the side of the well for 20 minutes.
- Remove the fixative and add new fixative as in the previous step, repeat this step once more
- Finally, slowly remove the microscopic slide of the multiwell plate and put it on a slide. Air-dry the slides at room temperature. Check spreading under phase-contrast microscope.

3.2.4 Harvesting procedures for flasks

This protocol will be considered later, in the description for the one used to obtain chromosome preparations from cell lines, since the procedure is the same (Protocol 3.3.2).

3.2.5 Banding techniques

There are several possibilities for G-banding; we will refer here two of the ones widely used for chromosome analysis. Its implementation depends on the laboratory conditions and standardization. The difference between them is the use of a reagent that allows the degradation of chromosomal proteins (trypsin or HCl) and the dye (Wright or Giemsa).

Protocol 1

Materials

- 0,2 N HCl
- 2xSSC Buffer, prewarmed at 65° in a water bath
- Wright's stain
- Sorense Buffer Tampon
- Disposal plastic pipettes
- Coupling Glass

Procedure

- Heat the slides in an oven at 70°C for 24 hours
- Remove the slides from the oven, leave for cooling and add 1-2 ml 0.2 N HCl on each slide for 2 minutes, using a Pasteur pipette
- Remove the HCl and thoroughly rinse with distilled water, let dry.
- Place carefully the slides in 2xSSC buffer, preheated at 65°C for 4 minutes.
- Remove the plates from the buffer and wash thoroughly with distilled water, let dry.

The time in HCl and in buffer depends on the type of cell; these times have been standardized for certain cell lines. If you do not get a good banding, you should standardize the time of exposure to these reagents.

- Dye the slides by adding 1-2 ml of Wright's dye solution on each slide, for 3 minutes.

Wright's stain solution is prepared by mixing 1 ml of Wright's stain with 3 ml of Sorensen buffer; this amount should be enough to dye 2 or 3 slides.

- Remove the stain and wash thoroughly with distilled water.
- View under a microscope to evaluate the banding quality.
- Cover slides with coverslips and seal them with Entellan to protect and preserve the chromosome spreads.

Protocol 2

Materials

- Trypsin 0.25%
- 2xSSC Buffer, prewarmed at 60° in a water bath
- Giemsa stain
- Sorense Buffer tampon
- Disposal plastic pipettes
- Coupling Glass

Procedure

- Dehydrate the slides in an oven at 80°C for 4 hours
- Remove the slides from the oven and carefully, place the slides in 2xSSC buffer preheated at 60°C for 30 minutes
- Remove the slides from the buffer and wash thoroughly with distilled water, let dry.
- Introduce the slides in the coupling glass containing a cold solution of trypsin with water (1:1) for 5 seconds

The trypsin stock solution for G-banding is at a concentration of 0.25%. The working solution consists of a 1:1 mixture with cold distilled water. This solution must be kept at 4°C, at this temperature best results are obtained.

- Remove the slides from trypsin and wash thoroughly with distilled water, let dry.

The time in trypsin depends on the type of cell; these times have been standardized for certain cell lines. If you do not get a good banding, you should standardize the time of exposure to these reagents. For best results, trypsin should always be kept cold and plates should remain hot. Once hot slides are introduced into cold trypsin, thermal shock can deliver better results. The slides can stay warm if these are placed around a hot plate

- Dye the slides by adding 1-2 ml of dye solution on each slide, for 10 minutes.

Giemsa stain solution is prepared by mixing 0,2 ml of Giemsa stain, 0,2 ml of Sorensen buffer and 4,6 ml of distilled water. This quantity is enough to dye 1 or 2 slides.

- Remove the stain and wash thoroughly with distilled water.
- View under a microscope to evaluate banding quality.
- Cover slides with coverslips and seal them with Entellan to protect and preserve the chromosome spreads.

If the chromosome spreads present a weak staining, Wright's or Giemsa solution can be added again for 2 minutes (Figure 5ᵃ). If the bands are too light, it is suggested to reduce the time in HC or trypsin. It is also recommended to start banding only at a slide; this will delineate the conditions needed to obtain a good banding. A correct banding for chromosome analysis consists on light and dark bands, which are clearly defined and have a proper amount of color (Figure 5b).

Note: It is important to control humidity and temperature when carrying out the banding; if low temperatures and high humidity are present, it is difficult to obtain a good banding.

3.3 Cell culture methods for cancer cell lines

Cell lines must be transported to the laboratory on dry ice and optimal conditions of sterility. If possible, these should reach the laboratory frozen and be kept at -20°C. Established cell lines are generally obtained from sources such as the American Type Cell Collection (ATCC), which are well adapted to *in vitro* growth. The ATCC Cell Biology Collection is the most comprehensive and diverse of its kind in the world, consisting on over 3,600 cell lines from over 150 different species. Some of the cell lines offered by ATCC are listed in Table 2.

3.3.1 Cell culture

All the solutions and equipment that come into contact with cells must be sterile, and proper aseptic techniques must be used. It is recommended before beginning each procedure to leave the laminar flow chamber exposed to UV radiation for 15 min. All cell culture incubations are carried out in a humidified incubator at 37°C and 5% CO_2.

Materials

- Appropriate culture medium (RPMI 1640 or DMEM), containing 10% fetal bovine serum (FBS), antibiotic-antimycotic solution (1X) and L-glutamine (2 mM).
- 75-cm^2 tissue culture flasks
- 15-ml conical centrifuge tube
- 5-ml and 10-ml plastic pipettes

Procedure

- Careful sterile techniques should be developed. Once in the laboratory, the cell line should be thawed in a serological bath at 37°C, avoiding contact with the lid.
- In a laminar flow chamber, transfer rapidly the cell suspension to a 15-ml conical centrifuge tube containing 5-10 ml of medium with FBS.
- Centrifuge the cell suspension for 5 min at 1500 rpm and room temperature.
- Pour off the supernatant; resuspend the cell suspension in 10 ml of medium and transfer cells to a sterile 75-cm2 culture.
- Place flasks on their sides in an incubator at 37°C and 5% CO_2 and leave them for 48 hr.
- At the end of incubation, examine the culture using phase-contrast microscopy to assess the extent of cell adhesion and cell growth.
- Change the culture medium and discard the culture medium existing. Add 5 ml PBS (1x) (removes non-adherent cells), remove PBS and add 10 ml of complete medium. Return flasks to incubator.
- Check cultures daily to determine the extent of adherence, cell growth and doubling time. The change of medium in cell cultures should take place every 48 hours, as previously indicated.

a.

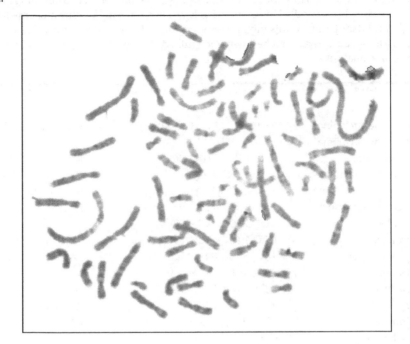

b.

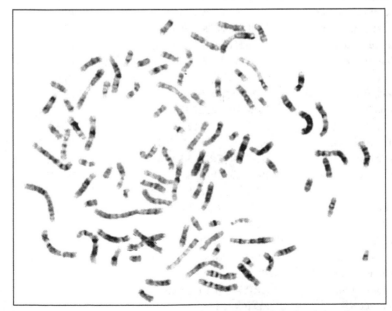

Fig. 5. G-Banded metaphase images (with and without correct banding) for chromosome analysis. (a) This metaphase shows very clear bands and homogeneously stained

chromosomes, the bands are not visible. In this case, the time of incubation in HCl or trypsin must be modified (reduced). This metaphase is not appropriate for chromosome analysis. (b) Light and dark bands in this metaphase are well defined and chromosomes have a proper staining; thus, these can be easily recognized. This metaphase is suitable for chromosome analysis.

Cell Type	Cell Name (ATCC® No.)	Media
Breast Adenocarcinoma	MDA-MB-231 (HTB-26)	Leibovitz's L-15
Breast Adenocarcinoma	MDA-MB-361 (HTB-27)	Leibovitz's L-15
Breast Adenocarcinoma	SKBR3 (HTB-30)	RPMI 1640
Breast Carcinoma	HCC1937 (CRL-2336)	RPMI-1640
Breast Adenocarcinoma	MCF-7 (HTB-22)	RPMI-1640
Breast Ductal Carcinoma	T47D (HTB-133)	RPMI-1640
Breast Ductal Carcinoma	BT474 (HTB-20)	DMEM
Colon Adenocarcinoma	COLO 205 (CCL-222)	RPMI-1640
Colon Cancer	DLD-1 (CCL-221)	RPMI-1640
Colon Carcinoma	T84 (CCL-248)	DMEM:F-12 Medium
Colon Carcinoma	CT26.WT (CRL-2638)	RPMI-1640
Cortical Neuron	HCN-1A (CRL-10442)	DMEM
Gastric Carcinoma	NCI-N87 (CRL-5822)	RPMI-1640
Hepatoma	Hepa 1-6 (CRL-1830)	DMEM
Kidney Fibroblast	COS-7 (CRL-1651)	DMEM
Lung Adenocarcinoma	NCI-H441 (HTB-174)	RPMI-1640
Lung Adenocarcinoma	NCI-H1975 (CRL-5908)	RPMI-1640
Lung Adenocarcinoma	NCI-H23 (CRL-5800)	RPMI-1640
Lung Carcinoma	NCI-H1299 (CRL-5803)	RPMI-1640
Lung Carcinoma	NCI-H460 (HTB-177)	RPMI-1640
Lung Carcinoma	NCI-H292 (CRL-1848)	RPMI-1640
Mammary Tumor	4T1 (CRL-2539)	RPMI-1640
Melanoma	B16-F10 (CRL-6475)	DMEM
Melanoma	A375 (CRL-1619)	DMEM
Pancreatic Beta Cells	Beta-TC-6 (CRL-11506)	DMEM
Pancreatic Cancer	AsPC-1 (CRL-1682)	RPMI-1640
Pancreatic Carcinoma	BxPC-3 (CRL-1687)	RPMI-1640
Pancreatic Carcinoma	MIA PaCa-2 (CRL-1420)	DMEM
Pancreatic Carcinoma	PANC-1 (CRL-1469)	DMEM
Prostate Cancer	VCaP (CRL-2876)	DMEM
Prostate Carcinoma	22Rv1 (CRL-2505)	RPMI-1640
Prostate Carcinoma	LNCaP clone FGC (CRL-1740)	RPMI-1640
Renal Adenocarcinoma	786-O (CRL-1932)	RPMI-1640
Retinal Epithelium	ARPE-19 (CRL-2302)	DMEM:F-12 Medium
Rhabdomyosarcoma	RD (CCL-136)	DMEM

Table 2. Some Human Cancer Cell lines offered by ATCC

- Once cell cultures reached the 80% of cell confluence, proceed with the protocol for the preparation of metaphase spreads

Note: Cells will be routinely monitored for mycoplasma contamination.

3.3.2 Preparation of metaphase spreads

In order to obtain metaphase spreads, cultures can be treated with Colcemid or Velbe (the use, exposure time and concentration of these arrest agents varies and depends on several factors, including cell type and overall growth characteristics), tripsinized, pelleted by centrifugation, hypotonically swollen and fixed. Incubation times depend on the type of cell. The hypotonically swollen and fixed cytogenetic suspension is then applied to glass slides and air-dried. To obtain good chromosomes spreading, the environment relative humidity should be of approximately 42%, with a temperature of 27°C. The slides are then ready for conventional or molecular cytogenetic analysis.

Materials

- Appropriate culture medium (RPMI 1640 or DMEM), containing 10% fetal bovine serum (FBS), antibiotic-antimycotic solution (1X) and L-glutamine (2 mM).
- 3,3 µg/ml Colcemid or 0,5 µg/ml Velbe
- Trypsin-EDTA (1x)
- PBS
- Hypotonic Solution: 0.075 M KCl or 20 mM KCl + 10 mM Sodium Citrate. Prewarmed to 37°C in oven or a water bath.
- 3:1 (v/v) methanol/acetic acid fixative
- 15-ml conical centrifuge tube
- Sterile disposable plastic transfer pipets
- 75-cm^2 tissue culture flasks
- Glass slides

Procedure

- Once cell cultures reached the 80% of cell confluence, observe under the microscope in order to determine the presence of dividing cells, this will ensure getting a good number of metaphases.

If the number of dividing cells is reduced, it is advisable to wait 24 hours before adding colchicine or Velbe. The dividing cells must be observed around (Figure 6).

- Add 20µl colcemid stock (3,3 µg/ml) for each 5 ml of culture medium to give a final concentration of 0,01µg/ml.
- Cap the flask securely and place it on its side in the CO_2 incubator at 37°C, continue incubating for 2.5 or 3 hours.

If cell growth is slow, it is recommended to use Velbe, in this case add 200µl of velbe solution (0,5 µg/ml) for each 10 ml of culture medium, to give a final concentration of 0,01µg/ml. Incubate for 16 hours.

- At end of incubation, transfer the medium to a 50-ml conical centrifuge tube and add on the culture flask 10 ml of sterile PBS; afterwards, remove carefully the PBS and transfer it to the same sterile 50-ml tube.

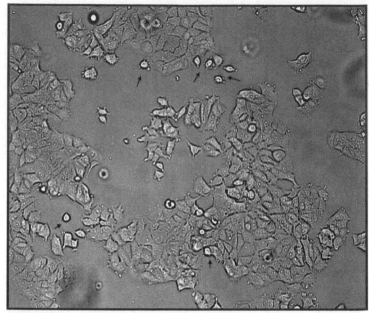

a. Cell line with few dividing cells (indicated by arrows)

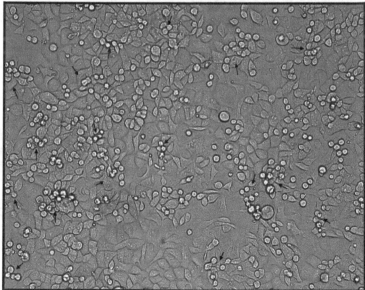

b. Cell lines with good number of dividing cells. Many groups of cells in division are observed (indicated by arrows)

Fig. 6. Cell culture displaying dividing cells. Figure a shows few dividing cells present, in this case, it is not recommended to add Colchicine or Velbe. Figure b shows good number of dividing cells, enough to add the arrest agents.

The addition of PBS to the culture flasks allows washing and removing medium traces that could interfere with the subsequent action of trypsin.

• Trypsinize attached cells in the flask using 1X Trypsin-EDTA and let stand for 3 minutes at room temperature.

Some cell lines need to be subjected to this step at 37°C, these temperature facilitates a fast cellular detachment, but again this will depend on the type of cell because at this temperature some cells can be damaged.

• When cells have detached, add fresh medium, collect cells and transfer them to the 50-ml tube containing the medium and the PBS.
• Centrifuge 10 min at 1500 rpm and room temperature. Discard the supernatant.
• Resuspend the cell pellet by gently tapping the tube base.
• Add to resuspended cells 2 or 3 ml of prewarmed (37°C) 0.075 M KCl (hypotonic solution) and mix gently using a plastic disposable transfer pipette. Incubate 15 min at 37°C.

It is also recommended to use the hypotonic solution formed by KCl 20 mM and Sodium Citrate 10 mM for 15 min at 37°C when longer, twisting or overlapping chromosomes are obtained.

• Add 2 ml of 3:1 cold methanol/acetic acid fixative. Cap tube and mix gently three times by inversion.
• Centrifuge for 10 min at 2500 rpm and room temperature.
• Discard supernatant and resuspend the pellet thoroughly by flicking the bottom of the tube.
• Add 3:1 cold methanol/acetic acid fixative and mix by performing continuous movements (60 times), using a Pasteur pipette
• Centrifuge again for 10 min at 2500 rpm and room temperature. Discard supernatant and resuspend pellet by flicking gently the tube bottom.
• Repeat fixation by adding 5 ml of fixative, resuspend 50 times with pipette.
• Place the tube into the refrigerator for 20 minutes at -20°C.

This step optimizes the action of the fixative solution and allows obtaining cleaner chromosome spreads.

• Centrifuge again for 10 min at 2500 rpm and room temperature. Discard supernatant and resuspend pellet by flicking gently the tube bottom.
• Repeat fixation and centrifugation once more.
• Discard fixative, add sufficient fixative in such a way that the suspension appears opaque and resuspend using a Pasteur pipette.
• Take a slide previously placed in ethanol, dry and clean it properly and hold roughly at a 45° angle.

Slides to be used for chromosomal analysis should be carefully cleaned and degreased; keep the slides in a container with 70% ethanol and stored at -20°C, this allows the slides to be clean and free of grease. Before its use, wipe them with a dust-free cloth.

• Using a 1-ml plastic disposable transfer pipet, add 3 drops of the cytogenetic suspension. Allow the slide to dry.

Is important to control or maintain proper temperature and humidity conditions (27°C, 42% humidity). This will allow us to obtain a proper chromosomal dispersion. If these conditions are not controlled, you will probably get closed metaphases with lapped chromosomes, which are not suitable for chromosome analysis.

- After laying the first slide, examine the following items by phase-contrast microscopy: mitotic index, chromosomal dispersion and presence of cellular debris. This will give an indication of the changes that need to be carried out in order to obtain optimal chromosome spreads.

If the chromosomal dispersion is not appropriate and you find lapped chromosomes, it is recommended to add distilled water on the slide and immediately drop the cell suspension (before the cell suspension). Afterwards, put the slide on serological bath preheated to 68°C, so that the steam produced baths the opposite side of the slide, where the chromosome spread has been performed. This will allow the slow evaporation of the fixative solution to contribute to good chromosome dispersion. If the evaporation of the fixative solution is fast, the chromosomes will not have enough time to separate from each other, and you may obtain lapped chromosomes (Figure 7). Otherwise, If the mitotic index is low, you could try applying Velbe (if colchicine has been previously applied). The low mitotic index could indicate a low proliferative index of the cell line (Figure 8)

- Store the pellet eventually left in an Eppendorf tube, using 1ml of fresh fixative at –20°C for further use.
- Finally, make banding following the protocol previously mentioned (Protocol 3.2.5)

The application of cytogenetics in cancer has acquired in the last two decades great importance not only as invaluable diagnostic tool but as a powerful research tool. As a diagnostic tool has allowed the identification of chromosomal aberrations and understanding among others, of malignant transformation in many cancers, which has provided important information about the biology of cancer. As a research tool provides

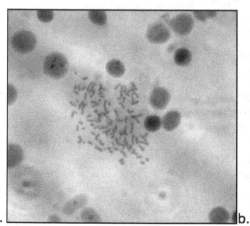
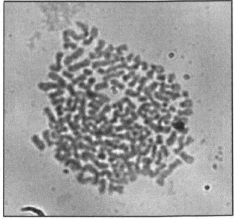

a.   b.

Fig. 7. Phase-contrast images of slides that are appropriate and not appropriate for cytogenetic analysis. (a) This metaphase shows good chromosome dispersion and chromosomes having a proper length for banding analysis. (b) This metaphase shows overlapping chromosomes, which are not appropriate for banding analysis.

a)

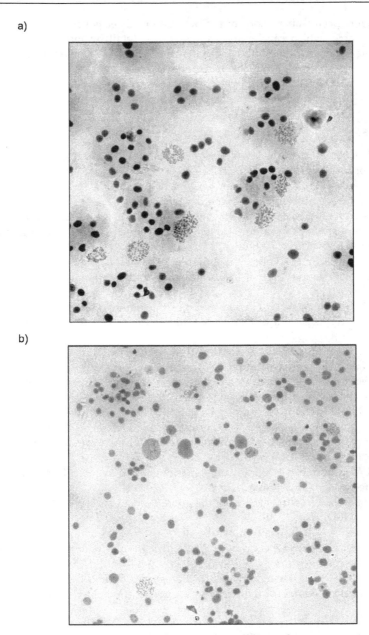

b)

Fig. 8. Phase-contrast images of slides that are appropriate and not appropriate for cytogenetic analysis. (a) Chromosome spread showing a good number of intact metaphases, which is appropriate for analysis and indicates that the dilution of the cell suspension in fixative is good. (b) Chromosome spread displaying a low number of metaphases, the cell suspension needs to be centrifuged again and resuspended in a smaller amount of fixative, which allows the concentration of a greater number of metaphases.

information about specific aberrations of a new type of cancer genes possibly involved in postulating the same, which can be analyzed in more detail by molecular studies, thus allowing a greater understanding of the molecular mechanisms envolved in carcinogenesis. Given the above, the knowledge and use of cytogenetic techniques allow proper application both in the diagnostic field or research directed toward improving our knowledge in various pathologies associated with chromosomal abnormalities.

4. References

Burdall, S. E., Hanby, A. M., Lansdown, M. R., & Speirs, V. (2003). Breast cancer cell lines: friend or foe? *Breast Cancer Res, 5*(2), 89-95.

Daly, E. M., & Taylor, R.E. (2009). Entropy and Enthalpy in the Activity of Tubulin-Based Antimitotic Agents. *Current Chemical Biology, 3,* 367-379.

Hevir, N., Trost, N., Debeljak, N., & Rizner, T. L. (2011). Expression of estrogen and progesterone receptors and estrogen metabolizing enzymes in different breast cancer cell lines. *Chem Biol Interact, 191*(1-3), 206-216.

Keydar, I., Chen, L., Karby, S., Weiss, F. R., Delarea, J., Radu, M., et al. (1979). Establishment and characterization of a cell line of human breast carcinoma origin. *Eur J Cancer, 15*(5), 659-670.

Lasfargues, E. Y., Coutinho, W. G., & Dion, A. S. (1979). A human breast tumor cell line (BT-474) that supports mouse mammary tumor virus replication. *In Vitro, 15*(9), 723-729.

Lasfargues, E. Y., Coutinho, W. G., & Redfield, E. S. (1978). Isolation of two human tumor epithelial cell lines from solid breast carcinomas. *J Natl Cancer Inst, 61*(4), 967-978.

Lattrich, C., Juhasz-Boess, I., Ortmann, O., & Treeck, O. (2008). Detection of an elevated HER2 expression in MCF-7 breast cancer cells overexpressing estrogen receptor beta1. *Oncol Rep, 19*(3), 811-817.

MacLeod, R. A., Dirks, W. G., Matsuo, Y., Kaufmann, M., Milch, H., & Drexler, H. G. (1999). Widespread intraspecies cross-contamination of human tumor cell lines arising at source. *Int J Cancer, 83*(4), 555-563.

Marcovic, O., & Marcovic, N. (1998). Cell cross-contamination in cell cultures: the silent and neglected danger. *In Vitro Cell Dev Biol, 34,* 108.

Masters, J. R. (2002). HeLa cells 50 years on: the good, the bad and the ugly. *Nat Rev Cancer, 2*(4), 315-319.

Masters, J. R., Thomson, J. A., Daly-Burns, B., Reid, Y. A., Dirks, W. G., Packer, P., et al. (2001). Short tandem repeat profiling provides an international reference standard for human cell lines. *Proc Natl Acad Sci U S A, 98*(14), 8012-8017.

Osborne, C. K., Hobbs, K., & Trent, J. M. (1987). Biological differences among MCF-7 human breast cancer cell lines from different laboratories. *Breast Cancer Res Treat, 9*(2), 111-121.

Van Bokhoven, A., Varella-Garcia, M., Korch, C., Hessels, D., & Miller, G. J. (2001). Widely used prostate carcinoma cell lines share common origins. *Prostate, 47*(1), 36-51.

Cytogenetic Analysis:
A New Era of Procedures and Precision

Diones Krinski[1], Anderson Fernandes[2,3]
and Marla Piumbini Rocha[4]
*[1]Federal University of Paraná,
Department of Zoology/Curitiba
[2]University of Mato Grosso State,
Department of Biology/Tangará da Serra
[3]Federal University of Viçosa,
Department of General Biology/Viçosa
[4]Federal University of Pelotas,
Department of Morphology/Pelotas
Brazil*

1. Introduction

The origin of cytogenetics is linked to technological advances. Cytogenetics differs from other sciences in which the observation of reality is sufficient to construct theories without the help of specialized equipment, as cytogenetics requires the use of imaging technology.

Theses technologies began with the discovery of the magnifying lenses, starting with microscopy and continuing into the era of informatics, allowing researchers to obtain more collaborative data to construct new hypotheses and substantiate theories. Novel ideas are constantly being developed, and science is continuously changing and reconstructing previous theories with new knowledge. In 1866, Gregor Mendel made his experiments with pea plants, using the existing technology of the time. He described the principles of genetics, but the technology did not permit the great scientist to infer the physical mechanisms involved in heredity. The function of chromosomes and hereditary was only understood half a century later, with the development of new technologies, improved microscopy for the study of cells, and new techniques to obtain chromosomes, which in turn led to the discovery of new methods of coloration and chromosomal banding. Cytogenetics arose from this fusion of knowledge.

Concurrent to these advances in cytogenetics, the field of molecular biology was expanding as well. Cytogenetics took advantage of this new knowledge and technology developed for molecular biology, thus forming a new area called molecular cytogenetics that is mainly represented by the techniques of FISH (*fluorescent in situ hybridization*) and microdissection.

The described improvements in the areas of genetics, cellular biology, cytogenetics and molecular biology were also accompanied by advances in the field of informatics. The creation of image analysis software permitted a refined analysis of banding patterns, making it easier to measure chromosomes and differentiate varying levels of coloration. With these new data, cytogenetics is now able to produce more specific responses about the composition of chromatin, the behavior of the chromosomes, karyotypic evolution, species phylogeny, as well as other aspects of molecular genetics.

1.1 A model of the advances in cytogenetics

Human cytogenetics has always had the greatest appeal in science, as is expected since man has always searched for his own self-knowledge. However, cytogenetics has also contributed to the generation of knowledge for other groups, one of which is the tribe of Meliponini bees. These bees are distributed in tropical climate regions where they are of great ecological and economic importance, due to their ability to pollinate cultivated native plants (Michener, 2000, Kerr et al., 1996; Heard, 1999).

Cytogenetic studies of native bees of the tribe Meliponini in Brazil began with the work of Kerr (1948), and until the 1980s the majority of the technical and applied works consisted of spread out metaphases and determining the number of chromosomes of each species. In 1988, Imai et al. published an article that used a new technique, called air drying, to obtain chromosomes with mitotic metaphases. That technique is now used in Meliponini cytogenetics and has shown itself to be the most efficient method of numbering and differentiating the chromosomes, in comparison to older spreading out techniques (Hoshiba & Imai, 1993; Brito et al., 2003; Mampumbu & Pompolo, 2000, Maffei et al., 2001, Rocha et al., 2003, 2007, Krinski et al., 2010, Barth et al., 2011, Lopes et al., 2011). For example, for the species *Melipona quinquefasciata,* individuals with chromosome numbers of n=9, 10 and 18 have been observed (Kerr, 1972; Tarelho, 1973). In 2002, Rocha et al. used the air drying technique to prove that the numerical difference was due to the presences of B chromosomes.

After the adaptation of the air drying technique, different methodologies were applied to the chromosomes of Meliponini. The most common of these techniques is C-Banding, following the BSG (Barium Hydroxide/Saline Solution/Giemsa) method of Sumner (1972), with modifications for different generas. This technique permits the differentiation of the euchromatin from the heterochromatin, as dark bands form in places with heterochromatin. These bands facilitated the identification of heteromorphism among homologues, the pairing of the chromosomes (Hoshiba & Imai, 1993) and assisted in the study of heterochromatin behavior during the mitotic cycle (Rocha et al., 2002).

Another banding technique is NOR (nucleolar organizer region) banding. Although the NOR is defined by a sequence of DNA, the NOR band is based not on the properties of DNA, but rather the acidic protein produced during interphase and accumulated in and around the NOR (Schwarzacher et al. 1978). In Meliponini, this technique was used according to Howell and Black (1980), making it possible to locate the NOR regions and use this data to study species evolution. Within molecular cytogenetics, various works were done using FISH and Restriction Enzymes (RE).

The use of the REs *in situ* is part of a new generation of cytogenetic techniques that employ the tools of molecular biology to provide responses about the molecular composition of chromatin (Pieczarka et al., 1998). As the name Restriction Enzymes implies, the DNA is cleaved in specific sequences. For example, HaeIII cleaves the sequence GC↓GC and DraI cleaves AAA↓TTT. According to Torre *et al.* (1993), the first work proposing a methodology for the use of the REs *in situ* for chromosomes fixed on slides was done by Mezzanotte *et al.*, in 1983. It showed that certain REs were capable of cleaving the fixed chromatin, removing the DNA in different dominions of the chromosome. The removal of the DNA from certain regions can be visualized by subsequent coloration with Giemsa or with fluorochromes, since the removal promotes the emergence of one or more transverse bands in the chromosomes or a simple reduction in the coloration of the regions (Verma & Babu, 1995). Rocha *et al.*, (2003a), using REs, showed that regions that appeared to be homogeneous in the C-Banding technique actually contained regions with different color intensities. The FISH technique uses dyes which are specifically connected to certain regions of the chromosomes and emit fluorescence. For example, Quinacrine Mustard (QM) and 4-6-diamidino2-phenylindole (DAPI) are connected to the regions rich in AT base pairs, while chromatin A_3 (CMA$_3$) is connected to the regions rich in GC. With the use of this technique, it is possible to infer the nature of the heterochromatin present in the karyotypes of various species, the origin of supernumery chromosomes and to study the nucleolar organizer region (Brito-Ribon *et al.*, 1998; Mampumbu & Pompolo, 2002, Brito *et al.*, 2003, 2005; Rocha *et al.*, 2003; Krinski *et al.*, 2010; Barth *et al.*, 2011).

More recently, Fernandes *et al.*, (2011) standardized a microdissection protocol for cytogenetic studies of bees. This new technique will permit the generation of extremely specific markings of the genome -- markings such as centromeric, telomeric, heterochromatic regions of supernumerary chromosomes, or even the entire chromosome itself. This technique has opened a window for the study of chromosomal evolution, and by consequence the genetic diversity of the species using a new method of analysis.

1.2 Image analysis in cytogenetics

Over the years, digital imaging technology has improved considerably, especially in studies related to microscopic analyses (such as cytogenetics) that essentially study the characteristics of the macro and micro chromosomal structures of the cells of diverse organisms. In recent decades cytogenetics has been transformed principally by developments in two areas of science, molecular biology and bioinformatics. The manipulation of DNA has been improved by the PCR (Polymerase Chain Reaction) technique, and advances in the FISH technique for chromosomal microdissection have taken chromosomal analysis to a new level of specificity and reliability in terms of studying karyotypical evolution, composition and rearrangements of species. The creation of image analysis software developed for use in bioinformatics has greatly enhanced the study of banding patterns, generating extremely reliable information with respect to chromosomal measurements and, with the use of coloration intensity, has enabled a comparison between these parameters in statistical form. The majority of the cytogenetic research currently done, whether focused on animals or plants, uses the various available software packages only to organize the chromosomes in order by size. The use of software for the study of images obtained by chromosomal banding techniques, and images of chromosomal and nuclear markings, is still not well known in cytogenetics. This is probably simply because most researchers do not know how these tools function. On the other hand, the increase in research using a wide variety of software makes these instruments become more sophisticated and precise each day.

In this chapter, we will discuss the basic use of the Image-Pro Plus™ (IPP) version 6.3 software program (Media Cybernetics, 2009) in the grouping and measurement of chromosomes, along with other methods of analyzing specific areas of interest, such as the macro and microstructures of chromosomes, which may be used in any cytogenetic study. As a model, we will use research done with karyotypes of stingless bees (Meliponini) in Brazil.

2. Using Image-Pro Plus™ for the study of chromosomes

2.1 Chromosome pairing and mounting of the karyotype

The majority of chromosomal studies start with the most basic technique of pairing the homologous chromosomes, as a pre-requisite for all other karyotype analysis. If this is done in a subjective manner, using only the visual perception of the researcher, then the pairing of the homologues may be inconsistent or doubtful, reaching a result that is not consistent with reality. Thus, to promote a more secure pairing of the chromosomes, the use of image analysis programs, such as IPP, is recommended. In this section we will show how these programs can act as powerful tools to be used in the most diverse studies of cytogenetics, so that the researcher can make a more detailed analysis in order to pair and mount the chromosomes in the correct manner.

Depending on the group of organisms being analyzed cytogenetically, some characteristics, such as the size of the chromosomes, the position of the centromeres and the distribution of the euchromatin/heterochromatin may be used to assist in the organization of the karyotype. To obtain a satisfactory result, this information must be taken into consideration when used in an image analysis program.

Not so long ago, the pairing of chromosomes meant that they had to be photographed, developed on photographic paper, cut out by hand with scissors, measured with tools (a caliper rule, for example), and then paired manually. This process was time-intensive and, as written above, gave subjective results most of the time.

With the use of the IPP program, we will demonstrate how technology can improve the accuracy of pairing chromosomes. Model chromosome images obtained from some species of wingless bees (Meliponini) will be used to perform the counting, separation and pairing of homologous chromosomes. However, before using these specific procedures, we emphasize that some precautions must be taken into consideration, including:

a.   The slides must be prepared in accordance with the standard procedures for cytogenetic analysis (conventional and technical specifications);
b.   The slides must be analyzed, and the best metaphases captured using a 100X objective and stored in a computer database. Remember that the higher the resolution of the image, the better the quality of the photographed metaphases;
c.   Your computer must have the IPP software (or similar) to analyze the images of chromosomes.

2.1.1 Counting the chromosomes

The IPP program has some tools which can be used to automatically determine the number of chromosomes of a given species. After the captured metaphase images are saved in a database on your computer, open the image that you wish to analyze and follow these instructions: go to the *Menu* bar and click *Measure* > *Count/Size* > *Select Colors*, and in the window marked *Segmentation*, click the *Color Cube Based* tab and select the *eyedropper* icon, then position the cursor over a chromosome and left-click to select the objects that have the same color. Note that the chromosomes will be partially selected (marked in red), repeat the procedure until all of the chromosomes are selected (Fig. 1 – step 1 and 2). After selecting all the chromosomes of the metaphase, close the *Segmentation*

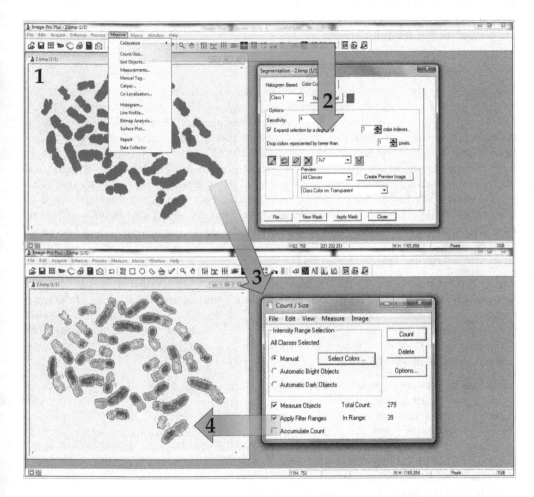

1- Chromosomes selected in red; 2 –*Segmentation* window in the *Color Cube Based* tab where the tools used to select the chromosomes that you wish to count are located; Automatic count of the chromosomes of the metaphase of *T. angustula fiebrigi*. 3- chromosomes numbered in green; 4–*Count/Size* window where the tools used to count the chromosomes will be available.

Fig. 1. Image-Pro Plus™ program with metaphase of *Tetragonisca angustula fiebrigi* (Schwarz, 1938) (Hymenoptera, Apidae, Meliponinae).

window and return to the *Count/Size* window. This time click on the button with the option to *Count*. Note that the chromosomes that were previously selected will now be numbered (Fig. 1 – step 3 and 4).

In the case of artifacts, such as nuclei or other structures with the same tonality as the chromosomes, these should be removed (deleted) from the figure with an image editing program first, otherwise they will also be selected and counted by IPP. In some cases, chromosomes may be very close, or even superimposed over one another. If this happens, the program may interpret these chromosomes as a single structure. To separate them, you should use another image editing program to copy the superimposed region to an area away from the other chromosomes.

2.1.2 Pairing homologous chromosomes

The pairing of the chromosomes is another important step for the mounting of the karyotype, because it allows the detection of homologues and possible alterations in the karyotype. In this way, chromosomal banding techniques have broadened the horizons of cytogenetics by showing specific regions, since each pair of chromosomes presents a distinct pattern and well-characterized bands. Therefore, the Q, G and C bandings are especially important in cytogenetics, because they permit the identification of small structural variations such as deletions, duplications and inversions, among other characteristics. This occurs because the bands show differences in the distribution of chromatin components or differences in the chemical composition of the chromatin along the chromosome, thus enabling the better understanding of chromosomal alterations established in each karyotype (Guerra, 1998). Here we will show the pairing done for the same metaphase used for the chromosomal count, using the C-banding technique.

After the previous steps, with the chromosomes still selected and already counted, return to the *Menu* bar and click *Measure > Sort Objects*. A new window and a dialog box of the same name will now be open. Select *Sort by: Area* and click on the box for *Sort Objects*. De-select the *Labels* option (the numbers that appear to the side of the chromosomes). The chromosomes will now be shown in a sequence, one that probably does not place the homologues side by side (Fig. 2).

Nevertheless, you can re-position them by clicking and dragging them. If you need to rotate a chromosome, just double-click on it to select it. Thus, you can organize them in what you think is the best possible way. After the chromosomes are organized in pairs, save the figure, preferably in "bmp" format. If any edits need to be made in the figure, such as inserting text and legends, you must be able to open the file in another image editing program (Corel Photo-Paint, Photoshop, etc) and make the necessary text insertions. If the chromosomes of the analyzed metaphase are very similar, we can use other tools to help in the pairing of the homologues, as shown in the next section.

2.1.3 Measuring chromosomes

The use of imaging software in the study of chromosomes generally makes the results more reliable, because through their use it possible to measure a curved chromosome simply by finding the two extremities of that chromosome (Barret & Carvalho, 2003). Here we will use

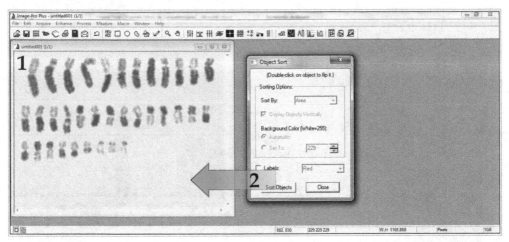

1 – Organized by area size. 2 - *Sort Objects* window where the options to choose the placement of the chromosomes are located.

Fig. 2. Chromosomes of *T. angustula fiebrigi* shown side by side.

IPP, as well as CorelDraw® and Microsoft Excel®, (similar programs may also be used) as tools to measure the chromosomes.

Thus, after capturing and processing the metaphases as explained in items 2.1.1 and 2.1.2, follow the following steps:

a. Save the image of IPP's automatic chromosomal pairings (Fig. 2). Next, import or open the figure in CorelDraw®;
b. In CorelDraw®, use the text tool to insert numbers below each chromosome and save the figure;
c. Open the numbered image in IPP to start the measurement of the chromosomes (Fig. 3 – step 1);
d. In the *Menu* bar, click *Measure > Measurements* and use the available tools to make measurements of the total size, the size of the arms, or the specific bands along the chromosomes (Fig. 3 – step 2);
e. Note that each measurement made will be numbered in the *Features* window. The measurements will be saved in IPP and can be used for calculating the Centromeric Index. These values should be exported to Microsoft Excel® with the *Input/Output* option (Fig. 3 – step 3). The image in IPP should also be saved;
f. The obtained values can be used to create graphs to organize the chromosomes by arm length or any other specific characteristic, such as the size of the euchromatin (Fig. 3 – step 4).

In figure 3, it is clear that IPP helped in the study of Lazaroto (2010), as this author used the obtained values in the graphs, especially the lengths of the euchromatic arms, to mount a karyotype (Fig. 3 – step 4) that presented more harmonious pairings among homologues. Note that the numbering in the final pairing does not correspond with the original numbering of the chromosomes. The problems presented in the first pairing were solved with the measure of each chromosome, and we can see the heteromorphism in homologous

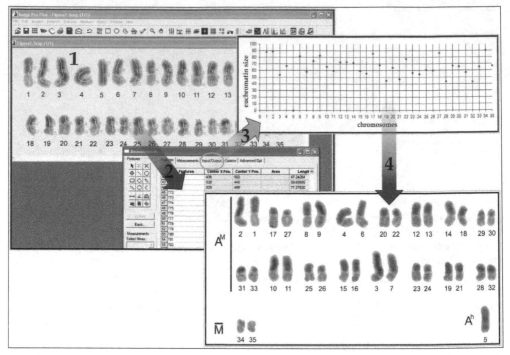

1 – Measurement and arrangement of chromosomes done without the use of IPP; 2 – Use of IPP to measure chromosomes and help in the grouping of homologues; 3 – Graphic made using data exported (*Input/Output*) from IPP to Microsoft Excel®. The points of each line indicate the size of the euchromatin; and 4 – Re-structured karyotype after the use of IPP. Observe the difference between 1 and 4.

Fig. 3. Karyotype of stingless bee *T. angustula fiebrigi*, previously numbered in CorelDraw®.

pairs among the heterochromatic arms of chromosomes 14 and 18, a feature that had not yet been described in the literature.

2.1.4 Important information in the mounting and pairing of chromosomes

1. Metaphase selection – The greater the difference between the tone of the background and the metaphase, the greater the quality of the karyotype. There are ways to amplify these differences both before and after image capture;
2. Follow the protocols precisely to obtain the best results with metaphases. Upon metaphase capture, the contrast between the chromosomes and the background can be increased by increasing the quantity of light emitted. After capture, images of metaphases can also be altered in image editing programs to achieve this contrast;
3. *Avoid metaphases containing overlapping*. There are ways to remove superimposed chromosomes, but this procedure requires time, ability, and the manipulation of reality. If you do not have the option of avoiding metaphases of this type, use programs such as CorelPhoto, Photoshop, etc.;
4. *NEVER alter your original figure*. It is not possible to restore the original figure after image-altering procedures. Always save a backup copy of your figures.

IMPORTANT

Procedures 3 and 4 imply alteration of the images, but the use of this method may cause the researcher to cross the line that separates research from scientific fraud.

2.2 Analyzing specific areas of chromosomes and nuclei

After knowing the karyotype of a species, the study of specific areas of the chromosomes is one of the principal objectives of cytogenetics, because these areas can reveal much information about the genetic evolution of the species or population studied. Thus, there are innumerable techniques to examine specific areas, such as Ag-NOR banding which shows the active nucleolar organizer regions (NORs), C-banding which shows the regions rich in heterochromatin, diverse types of fluorochromes that highlight the regions in specific base pairs, plus the techniques of FISH and Restriction Enzymes (REs). Here we will show how the IPP program can assist in the analysis of specific areas of nuclei and chromosomes.

Before showing the steps that must be followed to analyze specific areas of nuclei and chromosomes, we emphasize that the precautions described at the beginning of topic 2.1 must be followed. When the images are saved in the database on your computer and ready for analysis, follow the following steps.

2.2.1 Measuring areas of interest in nuclei and/or chromosomes

When you already have the images of the nuclei or chromosomes saved in a database on your computer, open the image that you wish to analyze by clicking *Menu > Measure*, and choose the *Measurements* option. A tab of the same name will open, so select the polygon icon and circle the region of which you wish to measure the area size (Figure 4 – step 3). Do not forget to select the area option, located on the left side of the *Measurements* window, below the options buttons (Select Meas:). Here, images of nuclei and nucleoli of the stingless bee species *Frieseomelitta trichocerata* Moure, 1990 (Fernandes *et al.*, 2008) and *Tetragonisca fiebrigi* Schwarz, 1938 are used to exemplify this process.

To see the values, and edit the name of each measured region, open the *Features* tab. To use the values obtained in other program just go to the *Input/Output* tab and choose the desired option (Fig. 4).

In cytogenetics, studies involving Ag-NOR banding, and more precisely those that refer to the relative size of the nucleoli found in interphasic nuclei, are scarce. Thus, the use of the tools of IPP by Fernandes *et al.*, (2008) provided information with respect to possible differences in the metabolic activities of cells of this species of bee. For both *F. trichocerata* and *T. fiebrigi*, the nuclei with three nucleoli (group III) presented a total area of nucleoli larger than that of the nuclei that presented only one nucleolus (group I), with this being a characteristic conserved among these species (Fig. 5). Therefore, these authors concluded that the nucleoli are cellular components which can offer important information about the metabolic activities and evolution of different organisms also indicate the technique of Ag-NOR banding as a possible cytological marker for distinct cellular types. We note that this tool can also be used to measure specific areas of chromosomes.

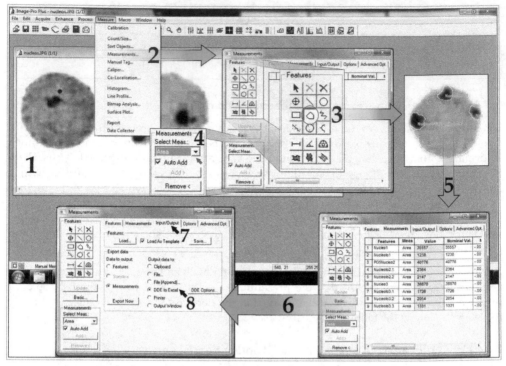

1 – Image open to start the area measurements by IPP; 2 – *Menu > Measure* with the *Measurements* option; 3 – Tools of the *Measurements* window, with the polygon (red circle), used to make measurements of the highlighted areas; 4 – Select the area option; 5 – Perform the measurements with the values shown in the *Measurements* tab; 6,7,8 –*Input/Output* tab where the values obtained in IPP can be exported to Microsoft Excel®.

Fig. 4. The IPP program open with nuclei and nucleoli of the stingless bee *T. fiebrigi* shown by the technique of Ag-NOR and previously numbered in CorelDraw®.

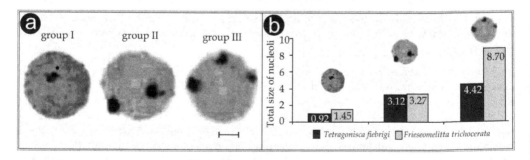

a) Three groups of nucleoli found in the nuclei of two species; and b) Differences in size (area) of the nucleoli of the nuclei of the analyzed species. Bar: 5 µm.

Fig. 5. Analysis of an area of nuclei and nucleoli of stingless bees *F. trichocerata* and *T. fiebrigi*.

2.3 Understanding chromosomal differentiation through the computational analysis of images

Historically, chromosomal banding techniques have assisted in trivial cytogenetic practices (such as the pairing of homologues in a more reliable manner) and more advanced analysis including understanding of patterns associated with chromosomal evolution of the species, the speciation processes, and the generation of genetic variability. Thus, we will discuss some research done with chromosomes of stingless bees that used the tools of IPP to assist in the cytogenetic analysis of this group of insects.

We will next show a re-analysis of the data obtained from the chromosomes of three species of the genus *Melipona* (*M. bicolor*, *M. quadrifasciata* and *M. subnitida*) which were subjected to digestion by two Restriction Enzymes (*Hae*III and *Dra*I) and colored with two fluorochromes (DAPI and CMA$_3$) (Fernandes, 2004). But first we need to understand what the Restriction Enzymes (REs) are and how they act on the chromosomes.

Thus, in this re-analysis we will use the results found after the degradation of the chromosomes of the three species of bees of the genus *Melipona* with the Restriction Enzymes (*Hae*III and *Dra*I), followed by coloration with the two fluorochromes (DAPI and CMA$_3$). In the first analysis done by Fernandes (2004), it was possible to verify the emergence of a complex banding pattern along of the chromosomes where a total of 135 were counted using the traditional visual counting method. Based on the similar and divergent banding patterns among these species, the author concluded that *M. quadrifasciata* and *M. bicolor* share a large number of bands in common when compared to the species *M. subnitida*.

However, the analysis of among-species band similarity caused some errors, since band equivalence was determined using only visual perception. A new proposal of band equivalency may be done using image analysis tools, which increases the spatial reliability (location) of the bands. In this analysis, IPP verified the exact locations of decreasing luminescence, generating a graph that indicates exactly where enzyme digestion occurs. Also, the overlap of the graphs allows precise comparisons between the bands, providing solid criteria for the conclusion about whether or not the bands within the chromosomes of the species are equivalent.

As an example of this analysis proposal, we used only the first chromosome of each species and treated it with *Hae*III, followed by coloring with DAPI (Fig. 6a). An extremely powerful tool of IPP involves the quantification of the light intensity of each pixel below a specific target (as seen in item 2.2.2). We then trace a line along the chromosome and the program quantifies the scale of shades of gray from 0 to 255 (from black to white), recording all of the pixels below this line (Fig. 6a). After this step, a histogram (graphic of pixel intensity by unit of length) can be made for each chromosome (Fig. 6b). In this way we will have the exact location of each chromosome degradation, and the overlap of the graphs provides the possibility of an extremely precise comparison between the patterns of digestion (Fig. 6c). Note that the valleys are highlighted by different geometric figures (triangles, stars and circles) for each chromosome.

Using this new analysis proposal, we note that there are three points at which there are high proximity between the chromosomal digestion of the three species (grouped by the pink elliptical circles). These locations of digestion have extreme proximity, and are the only unique points where we could consider that there is band similarity among the species.

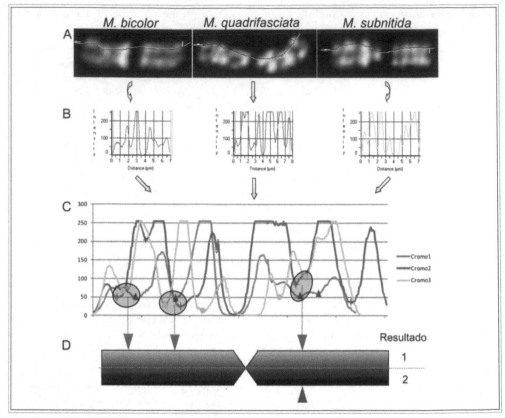

a) Arrangement of one chromosome of the first pairs from *Melipona bicolor, M. quadrifasciata* and *M. subnitida*, with the line above each chromosome indicating which pixels will be quantified to generate the histogram. b) Individual histograms of each chromosome. Note that in the graphic, the peaks represent the fluorescent regions and the valleys represent the regions of digestion caused by Restriction Enzymes. c) Overlap of the individual histograms. The regions of enzyme digestion are highlighted by different geometric figures (triangles, stars and circles) for each chromosome. The pink elliptical circle indicates locations of digestion present in the three chromosomes. d) The ideogram in which the exact locations of enzyme digestion in the three chromosomes are highlighted in accordance with the current proposal of diagnostic image analysis and by the previous proposal of Fernandes (2004) in which only traditional visual resources were used.

Fig. 6. Scheme of the procedures for the validation of the resulting bands of the restriction enzyme digestion (*Hae*III) followed by coloration with the fluorochrome DAPI.

Figure 1D represents the ideogram of a chromosome. This ideogram highlights the degradations of the enzymes, which we present as a new proposal of band similarity after image analysis in IPP in Result 1. The original proposal of Fernandes (2004) is shown in Result 2. We found that a more refined analysis allowed the visualization of two other digestion locations shared between the three species that, by the methods of traditional visualization, could not be distinguished, so it is possible that the conclusions of Fernandes (2004) can be modified.

Another example of the use of software analysis was recently presented by Krinski *et al.*, (2010). In this work, the authors described the karyotype of a stingless bee, *Oxytrigona* cf. *flaveola*. With the advent of image analysis, any of the various characteristics presented in karyotype may be evaluated. The first pair of chromosomes of this bee presents a heteromorphism of size, and in accordance with the application of fluorochrome CMA_3, this heteromorphism is due to a block of heterochromatin rich in CG bases. A thorough analysis with the help of IPP enabled the visualization of a subtle secondary constriction along the heterochromatic arm of the largest chromosome (Fig. 7).

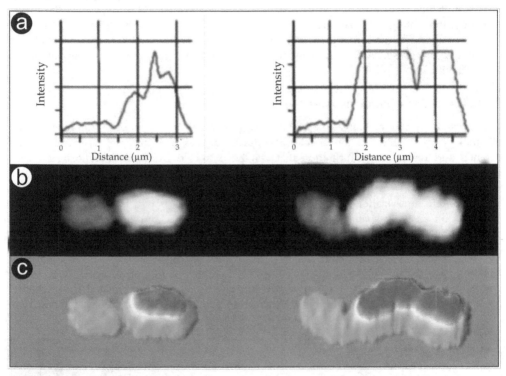

a) Histogram obtained from the fluorescence emitted from the chromosomes treated by fluorochrome CMA_3. b) Chromosomes treated with CMA_3. c) High relief image of the fluorescent chromosomes. Note that in a) and c) it is easy to see the secondary constriction present in the larger arm of the second chromosome, which is not possible in B). Figure courtesy of magazine Genetics and Molecular Biology (2010, Vol.33, No.3).

Fig. 7. Highlights the size heteromorphism presented by the first pair of chromosomes of the stingless bee *Oxytrigona* cf. *flaveola* (Krinski *et al.*, 2010,).

From the image obtained by the use of fluorochrome CMA_3, a graph was constructed which presented the fluorescence intensity of the pixels along the chromosomes. In this graph, secondary constriction was evident in the chromosome of the greatest length, represented by the valley in the graph. This constriction is located 3.5 micrometers from the start of the chromosome (see graphic in Fig. 7a) and causes a decrease of approximately 46% of the fluorescence in this location. It is here that image analysis shows its strength, since even a decrease of this magnitude could not be perceived using conventional optical analysis, as shown in the image of the chromosome.

In the same mounting, another resource that brought a refinement to the approach of the problem was the use of a topographic image of the chromosomes (Fig. 7c). This technique is based on the different light intensities emitted by those portions of the image (in this case by the chromosomes). Note that the smaller arms of both chromosomes present less fluorescence (reflection of the small quantities of the CG bases) which translates to a region of some increase in relief representation (Fig. 7c). The path of the long arm shows that the intensity of the fluorescence increases, and this increase is reflected in the rise of the image topology. Interestingly, the secondary constriction is visible using this resource, reinforcing the information obtained from the histogram. Either the relief or the graph may be analyzed to identify the presence of the secondary constriction, but only with the help of image analysis with IPP.

2.4 Linearization of chromosomes

Other resources may be suggested for better organization of cytogenetic data. In this way, the tool of image linearization enables a better visualization technique, which ultimately extends and refines the information that can be extracted from the chromosomes. Two commercial software packages can be used for this, *BandView*® (developed by Applied Spectral Imaging[Ltda]) or *Lucia Karyo*® (developed by Laboratory Imaging s.r.o. [Ltda]), both of which contain linearization tools . There are also free programs which can do chromosome linearization as well, such as *ImageJ* (Abràmoff *et al.*, 2004) and *Image SXM* (Barrett & Carvalho 2003).

In 2003, the karyotype of a species of bee (*Euglossa townsendi*) was presented by Fernandes *et al.* using various banding techniques. A suggested chromosomal pairing was presented by coloration with Giemsa. As an example, we submitted one of the suggested chromosomal pairs to the process of chromosomal linearization (Figure 8a). This process yielded new

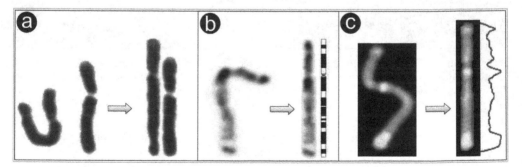

a) A pair of chromosomes colored with Giemsa that showed heteromorphism after the process. b) A chromosome subjected to the G-Banding technique, after image linearization it was possible to construct an ideogram of the banding patterns. c) Chromosome treated with fluorochrome CMA₃ the linearization enabled the construction of an extremely reliable graph of the luminescent intensity.

Fig. 8. Chromosome of bee *Euglossa townsendi* before and after the linearization of the images.

information, showing that the chromosomes were not of similar sizes, an observation that could not be made initially since one of the chromosomes was extremely curved. Now the karyotype of this species can be subjected to further analysis for us to discover if there is a size heteromorphism between the chromosomes of this pair, or if we have only a pairing error. The linearization of images can also be used for the construction of more precise ideograms (Figure 8b), and their use is not restricted to bright field microscopy, as it can also be used when the chromosomes are treated with fluorochromes. Note that independent of which proposal is used, the information obtained from straight chromosomes is larger in quantity and quality.

3. Conclusion

Note that in the most diverse areas of biological applications, and consequently in cytogenetic studies, image analysis through the use of current tools of bioinformatics, arises to overcome a series of limitations of common methods in the study of chromosomes. Considering this, and with the current technological developments in the fields of lenses, molecular biology and informatics, we can infer with more certainty about the mechanisms involved in karyotypic evolution. In a few years, other technologies are certain to be developed and new data will be obtained. Within this constant evolution of technology, the past cannot be ignored nor the current technology negated, because science is constructed on a base of pre-acquired knowledge, projecting itself to the future with the tools of the present moment

4. Acknowledgments

We are thankful for the financial support offered by the Fundações de Amparo à Pesquisa do Mato Grosso (FAPEMAT) and to the Conselho de Nacional de Desenvolvimento Científico e Tecnológico (CNPq). We also thank researchers Dr. Lucio Antonio de Oliveira Campos and Dr. Silvia da Graças Pompolo, who have dedicated a great part of their careers to the study of stingless bees.

5. References

Abràmoff, M.D.; Magalhães, P.J. & Ram, S.J. (2004). Image Processing with ImageJ. *Biophotonics International*, Vol.11, No. 7, pp. 36–42, ISSN 1081-8693.

Barrett, S.D. & Carvalho, C.R. (2003). A software tool to straighten curved chromosome images. *Chromosome Research*, Vol.11, pp. 83-88. ISSN 0967-3849.

Barth, A.; Fernandes, A.; Pompolo, S.G. & Costa, M.A. (2011). Occurrence of B chromosomes in *Tetragonisca* Latreille, 1811 (Hymenoptera, Apidae, Meliponini): a new contribution to the cytotaxonomy of the genus. *Genetics and Molecular Biology*, Vol.34, No.1, pp. 77-79, ISSN 1415-4757.

Brito, R.M.; Caixeiro, A.P.A.; Pompolo, S.G. & Azevedo, G.G. (2003). Cytogenetic data of *Partamona peckolti* (Hymenoptera, Apidae, Meliponini) by C banding and fluorochrome staining with DA/CMA3 and DA/DAPI. *Genetics and Molecular Biology*, Vol.26, No.1, pp. 53-58, ISSN 1415-4757.

Brito, R.M.; Pompolo, S.G.; Magalhaes, M.F.M.; Barros, E.G. & Sakamoto-Hojo, E.T. (2005) Cytogenetic characterization of two *Partamona* species (Hymenoptera, Apinae, Meliponini) by fluorochrome staining and localization of 18S r DNA clusters by FISH. *Cytologia*, Vol. 70, No.4, pp. 373-380, ISSN 0011-4545.

Brito-Ribon R.M.; Pompolo, S.G.; Martins, M.F.; Barros, E.G. & Sakamoto-Hojo E.T. (1998). Estudo da origem de cromossomos supranumerários em *Partamona helleri* (Hymenoptera, Apidae) por meio de hibridação in situ fluorescente e coloração com fluorocromos. *Anais do 3° Encontro sobre Abelhas* Vol.3, pp. 213–218.

Fernandes, A. & Pompolo, S.G. (2003). Comportamento da cromatina de *Euglossa townsendi* (Hymenoptera, Euglossinae) mediante as técnicas de bandas C, G e Enzimas de Restrição (HaeIII e SspI) seguidas de fluorocromos. Anais do 49° Congresso Nacional de Genética. Ribeirão Preto-SP : Sociedade Brasileira de Genética, 2003. p. 393.

Fernandes, A. (2004). Bandeamento cromossômico com Enzimas de Restrição e Fluorocromos no gênero *Melipona* (Hymenoptera: Apidae). Universidade Federal de Viçosa, Vicosa-MG, Brasil. Tese de Mestrado (Genética e Melhoramento), 43p.

Fernandes, A.; Barth, A.; Sampaio, W.M.S. (2008). BANDA Ag-NOR como marcador citológico e indicador da atividade sintética de células para duas espécies de abelhas sem-ferrão (Hymenoptera: Apidae: Meliponina). *Boletim de Pesquisa e Desenvolvimento (Embrapa Cerrados)*, Vol. 1, pp. 124.

Fernandes, A.; Scudeler, P.E.S.; Diniz., D.; Foresti, F.; Campos, L.A.O. & Lopes, D.M. (2011) Microdissection: a tool for bee genome studies. *Apidologie* (Celle), Vol. pp. , ISSN 0044-8435. (*in press*).

Guerra, M. (1988). *Introdução a Citogenética Geral*. Ed. Guanabara, ISBN 85-277-0065-4, Rio de Janeiro: Brazil.

Heard, T.A. (1999). The role of stingless bees in crop pollination. *Annual Review of Entomology*, Vol. 44, pp. 183–206. ISSN 0066-4170.

Hoshiba, H. & Imai, H.T. (1993). Chromosome evolution of bees and wasps (Hymenoptera, Apocrita) on the basis of C-banding pattern analyses. *Japanese Journal of Entomology*, Vol.61, pp. 465–492. ISSN 0915-5805.

Howell W.M. & Black D.A. (1980). Controlled silver-staining of nucleolus organizer regions with a protective colloidal developer: a 1-step method. *Experientia*, Vol.36, pp. 1014–1015. ISSN 0014-4754.

Imai, H.T.; Taylor, R.W.; Crosland, M.W.J. & Crozier, R.H. (1988). Modes of spontaneous evolution in ants with reference to the minimum interaction hypothesis. *The Japanese Journal of Genetics*, Vol.63, No.2, pp. 159–185, ISSN 1880-5787.

Kerr, W.E. & Silveira, Z.V. (1972). Karyotypic evolution of bees and corresponding taxonomic implications. *Evolution*, Vol.26, pp. 197–202, ISSN 1558-5646.

Kerr, W.E. (1948). Estudos sobre o gênero *Melipona*. *Anais da Escola Superior de Agricultura Luiz de Queiroz*, Vol.5, pp. 182–276, ISSN 0071-1276.

Kerr, W.E.; Carvalho, G.A. & Nascimento, V.A. (1996) *Abelha Uruçu – Biologia, Manejo e Conservação*. Fundação Acangaú, ISBN 978-546-6171-0-13, Belo Horizonte: Brazil.

Krinski, D.; Fernandes, A.; Rocha, M.P. & Pompolo, S.G. (2010) Karyotypic description of the stingless bee *Oxytrigona* cf. *flaveola* (Hymenoptera, Apidae, Meliponina) of a colony from Tangará da Serra, Mato Grosso State, Brazil. *Genetics and Molecular Biology*, Vol.33, No.3, pp. 494-498. ISSN 1415-4757.

Lazaroto, A.C. (2010). Aspectos citogenéticos de *tetragonisca fiebrigi* (Hymenoptera, Apidae, Meliponinae). Universidade do Estado de Mato Grosso. Tangará da Serra-MT, Brasil. Monografia (Ciências Biológicas), 21p.

Lopes, D.M.; Fernandes, A.; Praça M.; Werneck, H.; Resende, H.C. & Campos, L.A.O. (2011). Cytogenetics of three *Melipona* species (Hymenoptera, Apidae, Meliponini). *Sociobiology*, Vol.57, No ?, pp. 1-10. ISSN 0361-6525.

Maffei, E.M.D.; Pompolo, S.G.; Silva-Junior, J.C.; Caixeiro, A.P.; Rocha, M.P. & Dergam, J.A. (2001). Silver staining of nucleolar organizer regions (NORs) in some species of Hymenoptera (bees and parasitic wasp) and Coleoptera (lady-beetle). *Cytobios*, Vol.104, pp. 119–125. ISSN 0011-4529.

Mampumbu, A.R. & Pompolo, S.G. (2000). Localização da região organizadora de nucléolo por hibridização *in situ* na abelha sem ferrão *Friesella schrottkyi* (Friese, 1900) (Hymenoptera: Apidae: Meliponini), na região de Viçosa, Minas Gerais. Ribeirão Preto-São Paulo, pp. 1-723.

Mezzanotte, R.; Bianchi, U.; Vanni, R. & Ferrucci, L. (1983). Chromatin organization and restriction endonuclease activity on human metaphase chromosomes. *Cytogenetics and Cell Genetics*, Vol.36, pp. 562-566, ISSN 0301-0171.

Michener, C. D. (2000). *The bees of the world*. The Johns Hopkins Univ. Press, ISBN 978-080-1861-33-8, Maryland: USA.

Pieczarka, J.C.; Nagamachi, C.Y.; Muniz, J.A.P.C.; Barros, R.M.S. & Mattevi, M.S. (1998). Analysis of constitutive heterochromatin of Aotus (Cebidae, Primates) by restriction enzyme and fluorochrome bands, Chromosome *Research*, Vol.6, pp. 77–83, ISSN 0967-3849.

Rocha, M. P., Pompolo, S. G., Dergam, J.A.; Fernandes, A. & Campos, L.A.C. (2002). DNA characterization and karyotypic evolution in the bee genus *Melipona* (Hymenoptera, Meliponini). *Hereditas*, Vol.136, No. 4, pp. 19–27, ISSN 1601-5223.

Rocha, M.P.; Cruz, M.P.; Fernandes, A.; Waldschmidt, A.M.; Silva-Junior, J.C. & Pompolo, S.G. (2003). Longitudinal differentiation in *Melipona mandacaia* (Hymenoptera, Meliponini) chromosomes. *Hereditas*, Vol. 138, pp. 133-137, ISSN 1601-5223.

Rocha, M.P.; Pompolo, S.G.; Fernandes, A. & Campos, L.A.O. (2007). *Melipona* – Seis décadas de citogenética. *Bioscience Journal*, Vol.23, Supplement 1, pp. 111-117, ISSN 1516-3725

Schwarzacher, H.C.; Mikelsaar, A.V. & Schnedl, W. (1978). Nature of Ag-staining of nucleolus organizer regions. Electron microscopic and light microscopic studies on human cells in interphase, mitosis, and meiosis. *Cytogenetics and Cell Genetics*, Vol.20, No. 1, pp. 24-39, ISSN 0301-0171.

Sumner, A.T. (1972). A simple technique for demonstrating centromeric heterocromatin. *Experimental Cell Research*, Vol.75, pp. 304–306, ISSN 0014-4827.

Tarelho, Z.V.S. (1973). Contribuição ao estudo citogenético dos Apoidea. Dissertação (Mestrado em Genética) - Universidade de São Paulo, Ribeirão Preto.

Torre, J.; Lopez-Fernandez, C.; Gosalvez, J. & Bella, J.L. (1993). Restriction endonucleases: powerful tools to induce chromosome markers. *Biochemical Systematics and Ecology*, Vol. 21, No.1, pp. 13-24, ISSN 0305-1978.

Verma, R.S. & Babu, A. (1995). Human Chromosomes: Principles and Techniques. McGraw-Hill Inc., ISBN 978-007-1054-32-4. New York: USA.

Cytogenetics in Hematooncology

Ewa Mały[1], Jerzy Nowak[2]
and Danuta Januszkiewicz-Lewandowska[1,2,3]
[1]Department of Medical Diagnostics, Poznań
[2]Institute of Human Genetics Polish
Academy of Sciences, Poznań
[3]Department of Pediatric Hematology,
Oncology and Transplantology of
Medical University, Poznań
Poland

1. Introduction

Cytogenetic methods are widely used during diagnosis in many types of hematological malignancies. Classical methods like karyotyping using GTG banding technique and molecular methods like fluorescent *in situ* hybridization (FISH) are still gold standard in clinics all over the world. According to WHO 2008 classification the cytogenetic analysis is the basic diagnostic tool during the diagnosis of blood neoplasms. Chromosomal abnormalities have established and known prognostic significance. According to the European Leukemia Net-Workpackage Cytogenetics – the cytogenetic diagnostic is essential for disease classification, prognostic assessment and treatment decisions. Despite the fact that new molecular methods are very promising the classical cytogenetic methods together with FISH are still the reference test in many hematological neoplasms.

2. Cytogenetic techniques

Cytogenetic techniques are the basis in hematological diagnostics. From the first described chromosomal aberration characteristic for CML- Philadelphia chromosome by Nowel and Hungeford in 1960 it appeared that almost every hematological neoplasm possess typical changes in karyotype.

Bone marrow is the most suitable material for cytogenetic diagnostics in hematological malignancies. The most important principle is to aspirate the bone marrow in sterile way to heparin filled probes. Then the procedure of *in vitro* culture and preparing microscopic slides with metaphase chromosomes and/or interphase nuclei is started.

2.1 Classical cytogenetics

Chromosomes in metaphase are indispensable for karyotyping. For this purpose aspirated bone marrow cells are cultured *in vitro*. Preliminary preparation comprise the lysis in NH_4Cl solution to obtain mononuclear cells, which are then counted. These two important steps have a serious influence on further results on *in vitro* culture and finally on chromosome quality. Counted cells are placed in the medium with fetal bovine serum and appropriate growth factor. Non stimulated cells should be also parallel culture. Usually the culture takes 24 hours, but if the number of cells is sufficient, additional cultures for 48 hours or even 72 hours should be made. The most suitable for authentic diagnostics is 24-hours culture without any stimulating factor (only medium and serum).

In hematological patients especially pediatric, the aspiration of bone marrow could be problematic. Sometimes the quality of aspirated bone marrow is very poor, so it is very important in such cases to put more carefulness and attention during procedures of culture. After 23 hours of culture the cell cycle is stopped by incubation for 1 hour with colcemid solution which blocked the cells in metaphase. Then the procedure of hypotonic shock with KCl and Carnoy's fixation is begun to obtain chromosomes slides which are ready for banding techniques- the most common in the world- GTG banding using Giemsa stain and trypsin. Analysis is conducted using high quality microscope with special karyotyping software by experienced cytogenetists. The karyotype analysis covers the counting of the patient's chromosomes and structure study of each pair to detect any aberrations. The changes in the karyotype are described using international nomenclature with comments about eventually significance on prognosis.

2.2 Fluorescent *in situ* hybridization

The most important principle during preparation of slides for fluorescent *in situ* hybridization (FISH) is to obtain interphase nuclei from non-cultured and non-stimulated cells immediately after aspiration of material. Further differences between percentage of neoplasm cells in FISH analysis and in karyotype, result from fact that FISH technique is conducted on fresh material and karyotype after *in vitro* culture. So FISH represents the state of patient at the time of bone marrow aspiration. It could be possible that cells with aberration visualized by FISH are not detectable in karyotype. To obtain appropriate slides for FISH, cells after lysis, are incubated with KCl and fixated with methanol and glacial acetic acid.

Nowadays there is a wide spectrum of accessible commercial FISH probes ready for use created for almost every described chromosomal aberration observed in hematological malignancies. These commercial probes included translocation probe, deletion probe, centromeres, telomeres, painting probes (whole or only some fragments of chromosomes) available in many colors depending on possessed fluorescence filters in microscope.

3. Hematological malignancies

Hematological malignancies are a group of neoplasms derived from malignant transformed bone marrow cells. These types of cancers affect the peripheral blood, bone marrow, liver, spleen and lymph nodes. The great diversity in this group of disorders is due to the complexity in hematopoiesis and immunological system. Many classifications were created but the last most accepted is the World Health Organization (WHO) classification (2008) made by many hematologists from all over the world. The primary basis of this classification is the distinction of the origin of tumor cells: lymphocytes or myeloid cells.

Since chromosomal aberrations are common cause in the hematological malignancies and hence it is necessary to use the cytogenetic methods during the diagnosis and treatment evaluation.

3.1 CML – Chronic myeloid leukemia

Chronic myeloid leukemia (CML) is a clonal bone marrow stem cell disorder in which proliferation of mature granulocytes and their precursors is increased. The BCR/ABL fusion resulting from translocation t(9;22) is the well-known pathogenic key of this disease. The standard analysis during diagnosis cover the karyotype, FISH and molecular establishment of the type and the level of transcripts. Using the cytogenetic methods is very important during recognition of CML.

Usually in CML patient we can observe in the karyotype the translocation t(9;22) resulting as chromosome Philadelphia. About 1% cases of CML have so called masked chromosome Philadelphia (Virgili et al., 2008). Then in karyotype there is a normal structure of chromosomes (Fig. 1). In such cases it is obligatory to perform the FISH analysis to visualize the fusion of ABL and BCR genes (Fig. 2).

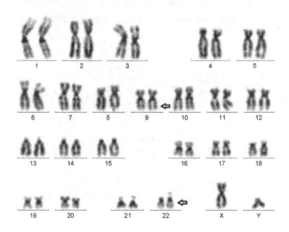

Fig. 1. CML patient with masked Philadelphia chromosome. Normal GTG banding karyotype shows no abnormalities between chromosomes 9 and 22.

Apart from classical translocation t(9;22) in karyotype we can observe different secondary aberrations. We had a patient with three aberrations in the karyotype during blast crisis: t(9;22) with BCR/ABL fusion, t(3;21) with EVI/AML1 fusion and an inversion of chromosome 2. Knowledge of the importance of specific translocations, like t(3;21), can help clinicians. Our patient had no hematological and clinical sings of relapse, but appearance of t(3;21) translocation indicates the clinicians about the upcoming progression of the disease (Kim et a., 2009). It could not be possible to detect this changes using only FISH method or RT-PCR. Only karyotype could reveal all the three aberrations. (Fig.3).

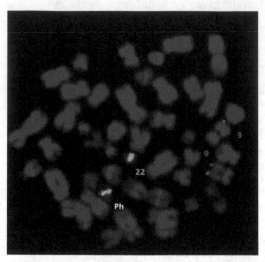

Fig. 2. CML patient with masked Philadelphia chromosome. FISH analysis revealed the fusion between BCR and ABL genes on chromosome 22. The probe BCR/ABL dual color, dual fusion- red signals seen on chromosome 9q34, green- on chromosome 22q11.2.

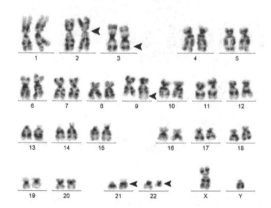

Fig. 3. Karyotype of patient with CML revealed three aberrations: t(9;22)(q34;q11), t(3;21)(q26;q22) and inversion of chromosome 2.

3.2 AML – Acute myeloid leukemia

Acute myeloid leukemia (AML) is a clonal disorder characterized by various genetic abnormalities and variable response to the treatment. In AML, cytogenetic methods are used to stratify patients to three different risk groups: good, intermediate and poor (Akagi et al, 2009). Karyotype feature in the favorable risk group include t(8;21), t(15;17) and inv(16)/t(16;16). In the intermediate risk group characteristic feature of karyotype are: normal karyotype, -Y, del7q, del9q, t(9;11), del11q, isolated trisomy 8, +11,+13,+21, del20q. Aberrations with adverse prognosis are: complex karyotype, inv(3)/t(3;3), monosomy 5, 7, t(6;9), t(6;11), t(11;19). Unfortunately about 40-50% of AML cases represent normal karyotype with intermediate prognosis (Rausei-Mills et al, 2008). Due to this fact additional molecular genetic investigations are of increasing importance during each new diagnosed AML case.

For patients with *de novo* AML the best clinical approach is the classical cytogenetic analysis including FISH technique to categorize patients into specific risk group. In AML patients with normal karyotype it is strongly recommended to examine molecular background of the disease -detection of mutations in genes such as FLT3, NPM1 or CEBPA (Döhner et al, 2008).

Although translocation t(15;17) with fusion of PML/RARα genes characteristic for acute promyelocytic leukemia (APL) has established good prognosis, in some cases it could not be so easy to interpret. In our laboratory there was a patient (2-years-old boy) with diagnosed AML-M3. FISH analysis revealed single fusion of PML/RARα genes (Fig.4), with three signals from PML gene (green), and two from RARα gene (red). Karyotype was very difficult to analyze, and many FISH procedures with different probes were needed to established at least very complex karyotype with PML/RARα fusion gene on derivative chromosome 16 (Fig. 5). ISCN record of karyotype showed as follow:

46,XY,der(4)(4pter→4q26::16q22→16qter),der(15)(15pter→15q23::4q31.3→4qter),der(16)(16q22→16p11.2::17q25→17q21::4q26→4q31.3::17q21::15q23→15qter),der(17)(17pter→17q21::16p11.2→16pter) [20]

Schematic representation (Fig.6) indicates possible way of process of occurrence of aberrations. Red color represents the chromosome 4, blue- 15, green- 16 and grey- 17. The fusion of PML/RARα genes on derivative chromosome 16 was also indicated. So complex karyotype shows rather poor prognosis in this patient.

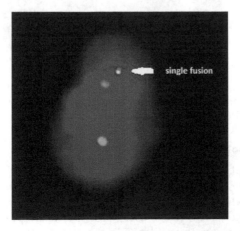

Fig. 4. FISH with PML/RARα probe revealed single fusion (indicated by arrow).

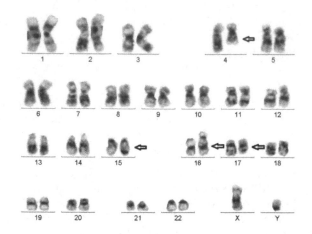

Fig. 5. Karyotype of APL patient with complex rearrangements involving chromosomes 4,15,16 and 17.

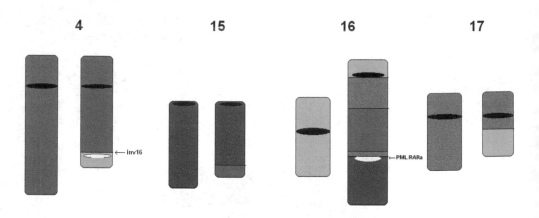

Fig. 6. Schematic representation of complex aberrations involving chromosome 4,15,16 and 17 in the karyotype of APL patient.

3.3 ALL – Acute lymphoblastic leukemia

Acute lymphoblastic leukemia (ALL) is the most often childhood neoplasm account about 30% of all cancer. Frequency of ALL depends on the age and in children estimates about 85% acute leukemias, while in adults only about 20%. The highest morbidity in children concern age between 2 and 6 (Deangelo, 2005). ALL is a heterogeneous group of leukemias which can differ between patients depending on age. The characteristic feature is accumulation of immature lymphoblasts in the bone marrow, peripheral blood and lymph nodes (Hamouda et al., 2007). Now a huge progress in treatment results and long-term remission is noted in children with ALL. About 80% of patients achieved complete remission and 5-year event free survival (Brassesco et al., 2011). Despite these facts relapses are still the main reason of treatment failure.

The most important step during diagnosis of ALL is the assessment of biological and genetic features of lymphoblasts (Tucci and Arico, 2008). Usually the prognostic factors are: gender, age, WBC, cytogenetic and immunophenotypic features of bone marrow cells, steroid resistance and time of remission achievement. Based on these factors the risk groups are determined as: standard, intermediate and high (Boer et al., 2009).

The most frequent (25% children's ALL) chromosomal aberrations with good prognosis are translocation t(12;21) with ETV6/RUNX1 genes fusion and hiperdiploidy more than 50 chromosomes. From poor prognostic cytogenetic factors the translocation t(9;22), MLL rearrangements and hypodiploidy under 45 chromosomes are listed. Normal karyotype has the intermediate prognosis (Harrison et al., 2010). Characteristic features for ALL blast cells show short time survival and low proliferation ratio, so it is relatively often (about 16% of ALL cases), it is impossible to obtain metaphases and established karyotype (Heng et al., 2010).

We presented here the very rare aberration typical for B-ALL- the dicentric chromosome dic(9;20). After GTG banding (Fig. 7) this chromosome form was recognizable, but additional FISH with whole chromosome probes resolved the deletion or additional aberrations of chromosome 20 arm p (Fig. 8.). Median age for patients with this aberration is 3 years, and it more frequently occurred in girls. ALL patients with dic(9;20) responded well for initial therapy, but relapses are detected relatively often.

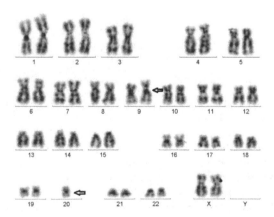

Fig. 7. The karyotype 45,XX,dic(9;20)(p13;q11),del(17)(q23),-20 of patient with ALL presenting the dicentric chromosome between chromosome 9 and 20.

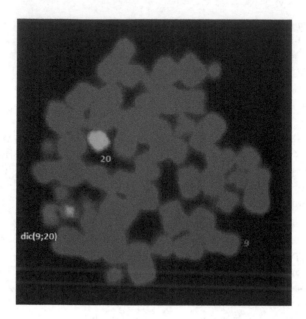

Fig. 8. FISH with whole chromosome probes for chromosome 9 (WCP 9 red) and 20 (WCP 20 green) in ALL patient with dicentric chromosome dic(9;20).

3.4 MDS – Myelodysplastic syndrome

Myelodysplastic syndromes are a heterogeneous group of clonal disorder of pluripotent hematopoietic stem cells which are characterized by ineffective hematopoiesis, peripheral blood cytopenia and fibrosis in the bone marrow. MDS often evolve to acute myeloid leukemia. Cytogenetic aberrations described in MDS patients, are common in AML as well (Vardiman et al., 2009). Myelodysplastic syndrome can be primary and secondary (therapy related) disorder. The most common difficulty in laboratory work with material from patients with MDS is cytopenia. It means, that there is very few cells obtained from clinicians to analyze. So careful proceeding with such a material, not to waste any cell, is the most important key.

We presented here one case of primary MDS with unbalanced recurring chromosomal abnormality isodicentric of chromosome X- idic(X)(q13) (fot.9) and one patient with therapy-related MDS with balanced translocation t(2;3) and monosomy 7 (fot.10,11).

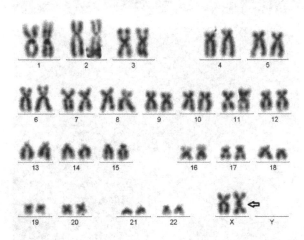

Fig. 9. Karyotype of female with primary MDS with characteristic double centromere in one of chromosome X - 46,XX,idic(X)(q13).

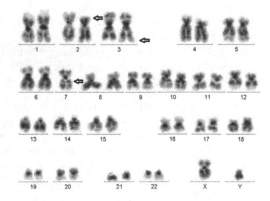

Fig. 10. Karyotype of patient with secondary MDS with balanced translocation between chromosomes 2 and 3 and monosomy 7- 45,XY,t(2;3)(p15;q27),-7.

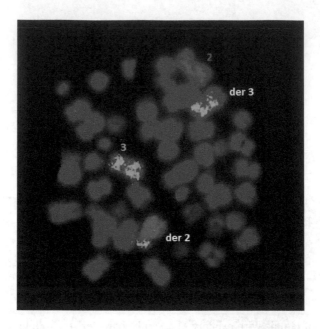

Fig. 11. FISH confirming the balanced translocation t(2;3)(p15;q27) in patient with secondary MDS.

3.5 CLL – Chronic lymphocytic leukemia

Chronic lymphocytic leukemia (CLL) is the most common type of leukemia in the Western world and primarily affects the adults. CLL is characterized by accumulation of immunologically incompetent B lymphocytes in the bone marrow, blood and lymphoid organs (Hodgson et al., 2011). CLL is a biologically heterogeneous illness, that in some patients has an indolent feature that not require any therapy, while in others run with aggressive leukemia, which need immediate treatment (Bryan and Borthakur, 2010).

Almost 80% patients with CLL have acquired chromosomal abnormalities. The most frequent is deletion 13q (55% of patients) with rather good prognosis. Deletion 11q, especially deletion of ATM gene is found in 18% of CLL patients and is associated with adverse prognosis. Trisomy of chromosome 12 and normal karyotype (16-18%) have no established risk prognosis, while deletion 17p (found in 7% of individuals) related with deletion of p53 gene has the worse prognosis and highest risk of the treatment failure (Döhner et al., 2000).

We presented here simple examples of non-deleted and deleted interphase nuclei from CLL patients after FISH technique (Fig.12 a,b). CLL is the only hematological disorder, in which reference material for cytogenetic analysis can be peripheral blood. Bone marrow aspiration from CLL patients is very rare, and usually the standard proceeding, including FISH analysis with 4 probes specific for deletion 13q14, ATM gene, p53 gene, and trisomy 12.

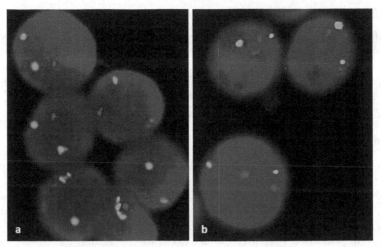

Fig. 12. a) FISH with p53 probe in patient presenting deletion of p53 gene (one red signal from p53 gene, two green signals from centromere of chromosome 17) (b). FISH with p53 probe on interphase nuclei of patient with CLL- no deletion of p53 gene was detected (with two red and two green signals).

Sometimes CLL patients examined only with FISH technique can be misled by the clinicians. We had an individual with known deletion 13q14 (rather good prognosis), while karyotype revealed an additional aberration - translocation t(7;14)(p15;q11),which is rather rarely observed, but characteristic for T-ALL (Fig. 13). This abnormality in CLL patient could indicate on possible progression toward acute lymphoblastic leukemia.

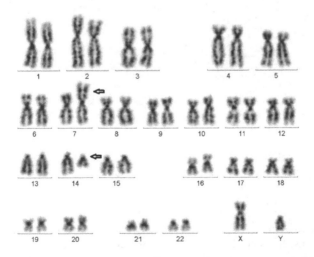

Fig. 13. CLL patient with deletion of 13q14 region (invisible in GTG banding- detected by FISH) and translocation between chromosomes 7 and 14- characteristic for T-ALL.

3.6 MPN – Myeloproliferative neoplasms

Myeloproliferative neoplasms are characterized by proliferation of multipotent stem blood cell. According to WHO (2008) these neoplasms can be divided into eight disorders not taking into account the presence the BCR/ABL transcript (Vannucchi et al., 2009). Except the chronic myeloid leukemia, all other myeloproliferative neoplasms are negative for BCR/ABL transcript, but the common feature is the frequent presence of JAK2 mutation. Myeloproliferative neoplasms are characterized by increased proliferation of one up to three hematopoietic cell lineages in the bone marrow, what has the great impact on peripheral blood parameters.

At present, after exclusion of CML, all patients are tested toward the JAK2 mutation and eventually in the direction of any other changes in karyotype. Cytogenetic aberrations are rather uncommon in MPN. In our laboratory however, there were few cases with specific and interesting changes in karyotype. Important thing to remember while working with material form MPN patients is accurate counting of mononuclear cells. Too large number of cells in the *in vitro* culture may result in failure to obtain the chromosome metaphase.

We present patient with MPN, which was positive for the JAK2 V617F mutation and has a duplication in the long arm of chromosome 1- a rare aberration in this type of hematological disorder (Fig. 14).

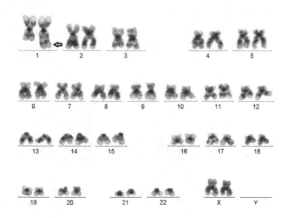

Fig. 14. MPN patient with duplication in the long arm of chromosome 1 - 46,XX,dup(1)(q21q32).

4. Conclusion

Acquired chromosomal aberrations are the leading molecular cause in the development of neoplasms. Deletions, translocations, amplifications which are clonal changes in karyotype can be detected in most of hematological disorders. Cytogenetic techniques like GTG karyotyping and FISH are excellent to reveal these aberrations. Also molecular techniques suitable for detection of mutations and quantitative measurement of known pathogenic transcripts have great importance. In molecular diagnostics of hematological malignancies in XXI century there is a great need to connect cytogenetic and molecular techniques. In our laboratory parallel GTG karyotype, FISH and molecular methods are carried out for diagnosis of hematological patients.

5. References

Akagi T., Ogawa S., Dugas M., Kawamata N., Yamamoto G., Nannya Y., Sanada M., Miller C.W., Yung A., Schnittger S., Haferlach T., Haferlach C.,&Koeffler HP. (2009). Frequent genomic abnormalities in acute myeloid leukemia/myelodysplastic syndrome with normal karyotype. *Haematologica*, Vol. 94, No. (2), pp:213-223.

Boer M.L., Slegtenhorst M., De Menezes R.X., Cheok M.H., Buijs-Gladdines J.G., Peters S.T., Van Zutven L.J., Beverloo H.B., Van der Spek P.J., Escherich G., HorstmannM.A., Janka-Schaub G.E., Kamps W.A., Evans W.E.,&Pieters R. (2009). A subtype of childhood acute lymphoblastic leukemia with poor treatment outcome: a genome-wide classification study. *Lancet Oncolology*, Vol. 10, No. 2, pp: 125-134.

Brassesco M.S., Xavier D.J., Camparoto M.L., Montaldi A.P., de Godoy P.R., Scrideli C.A., Tone L.G.,& Sakamoto-Hojo ET. (2011). Cytogenetic instability in childhood acute lymphoblastic leukemia survivors. *J Biomed Biotechnol*.article ID: 230481.

Bryan J.,&Borthakur G. (2010). Role of rituximab in first-line treatment of chronic lymphocytic leukemia.*Therapeutics and clinical risk management*, Vol. 22, No. 7, pp:1-11.

Deangelo DJ. (2005). The treatment of adolescents and young adults with acute lymphoblasticleukemia. *Hematology (Am SocHematolEduc Program):* 123–130.

Döhner H., Stilgenbauer S., Benner A., Leupolt E., Kröber A., Bullinger L., Döhner K., Bentz M.,&Lichter P. (2000). Genomic aberrations and survival in chronic lymphocytic leukemia.The New England journal of medicine, Vol. 343, No. 26, pp:1910-1916.

Döhner K.,&Döhner H. (2008). Molecular characterization of acute myeloid leukemia. *Haematologica*, Vol. 93, No. 7, pp: 976-982.

HamoudaF., El-Sissy A.H., Radwan A.K., Hussein H., Gadallah F.H., Al-Sharkawy N.,Sedhom E., Ebeid E.,& Salem S.I. (2007). Correlation of karyotype and immunophenotype in childhood acute lymphoblastic leukemia; experience at the National Cancer Institute, Cairo University, Egypt.*Journal of the Egyptian National Cancer Institute*, Vol. 19, No. 2, pp: 87-95.

Harrison C.J., Haas O., Harbott J., Biondi A., Stanulla M., Trka J., Izraeli S; Biology and Diagnosis Committee of International Berlin-Frankfürt-Münster study group (2010). Detection of prognostically relevant genetic abnormalities in childhood B-cell precursor acute lymphoblastic leukaemia: recommendations from the Biology and Diagnosis Committee of the International Berlin-Frankfürt-Münster study group.*British journal of haematology*, Vol. 151, No. 2, pp:132-42.

Heng J.L., Chen Y.C., Quah T.C., Liu T.C.,&Yeoh A.E. (2010). Dedicated cytogenetics factor is critical for improving karyotyping results for childhood leukaemias - experience in the National University Hospital, Singapore 1989-2006.Annals of the Academy of Medicine, Singapore, Vol. 39, No. 2, pp:102-106.

Hodgson K., Ferrer G., Montserrat E.,& Moreno C. (2011). Chronic lymphocytic leukemia and autoimmunity: a systematic review. *Haematologica*, Vol. 96, No. 5, pp:752-761

Kim T.D., Türkmen S., Schwarz M.,Koca G., Nogai H., Bommer C., Dörken B., Daniel P.,&le
    Coutre P. (2010). Impact of additional chromosomal aberrations and BCR-ABL
    kinasedomainmutations on theresponsetonilotinib in Philadelphia chromosome-
    positivechronicmyeloidleukemia. *Haematologica*, Vol. 95, No. 4, pp:582-588.
Rausei-Mills V., Chang K.L., Gaal K.K., Weiss L.M.,& Huang Q. (2008). Aberrant expression
    of CD7 in myeloblasts is highly associated with de novo acute myeloid leukemias
    with FLT3/ITD mutation.*American journal of clinical pathology*, Vol. 129, No. 4,
    pp:624-629.
Tucci F.,&Aricò M.(2008). Treatment of pediatric acute lymphoblastic leukemia.
    Haematologica, Vol. 93, No. 8, pp: 1124-1128.
Vannucchi A., Guglielmelli P.,&Tefferi A. (2009). Advances in understanding and
    management of myeloproliferative neoplasms.*CA: a cancer journal for clinicians*, Vol.
    59, pp:171-191.
Vardiman J.W., Thiele J., Arber D.A., Brunning R.D., Borowitz M.J., Porwit A., Harris N.L.,
    Le Beau M.M., Hellström-Lindberg E., Tefferi A.,& Bloomfield C.D. (2009). The
    2008 revision of the World Health Organization (WHO) classification of myeloid
    neoplasms and acute leukemia: rationale and important changes. *Blood*, Vol. 114,
    No. 5, pp:937-951.
Virgili A., Brazma D., Reid A.G., Howard-Reeves J., Valgañón M., Chanalaris A., DeMelo
    V.A., Marin D., Apperley J.F., Grace C.,&Nacheva E.P.(2008). FISH mapping of
    Philadelphia negative BCR/ABL positive CML. *Molecular Cytogenetics*, 1:14.

4

Microtechnologies Enable Cytogenetics

Dorota Kwasny, Indumathi Vedarethinam,
Pranjul Shah, Maria Dimaki and Winnie E. Svendsen
Technical University of Denmark, Department of Micro- and Nanotechnology
Denmark

1. Introduction

In this Chapter the standard cytogenetic methods are shortly introduced. Furthermore, the existing microtechnologies that improve the cytogenetic analysis are thoroughly described and discussed.

1.1 Traditional and molecular cytogenetics

Cytogenetic analysis is an important tool in pre- and postnatal diagnosis as well as cancer detection. In a traditional cytogenetic technique known as karyotyping the metaphase chromosome spreads are prepared on a glass slide and stained with a Giemsa stain. The stain reveals a specific banding pattern for each chromosome – a chromosome bar code. Karyotyping is often supplemented by the molecular cytogenetic technique Fluorescent *In Situ* hybridization (FISH), which requires the use of fluorescently labeled DNA probes to target a specific chromosome region. In FISH the chromosome preparations (metaphase spreads or interphase nuclei) are heat denatured, followed by application of the probe and hybridization at 37 °C. FISH can be performed on interphase nuclei on non-cultured cells in less than 24 hrs, but the chromosome structure cannot be visualized. On the other hand, metaphase FISH has the advantage of visualizing the entire karyotype at once and can detect potential abnormalities at a high resolution. But, the long analysis time and culturing required for metaphase FISH are important disadvantages.

Recently, a common DNA analysis, such as PCR amplification of a specific DNA region gained more popularity. Such analysis is beneficial as it can be performed on non-cultured cells, providing the results within a few days. Even though DNA techniques hinder the evolution of FISH, it can still provide valuable information on abnormalities, enabling detection of complex chromosome rearrangements. Nevertheless, FISH is now rarely used as the first step in cytogenetic analysis, due to the high cost of the probes, need for skilled technicians and lengthy analysis protocol. However, the use of microdevices for FISH could reestablish the status of this technique as an important tool for high resolution detection of chromosome abnormalities.

The major drawback of the FISH is the long analysis protocol. To perform a complete FISH analysis, even well trained technicians spend several hours in sample preparation as well as the waiting time in between each pre- and post- hybridization washes. There are at least 12-

15 different washes in a standard routine test that in total takes about 45 minutes. Apart from the cell culture work, the hybridization process is very time consuming. At minimum, performing a FISH analysis with centromeric probes (repetitive sequences), will take 2 to 3 hours. Furthermore, in some FISH experiments the hybridization of a probe requires overnight incubation. Another bottleneck of FISH analysis is the cost of the reagents used for the assays, mainly the fluorescent probes. In standard lab protocols 10-15 µl of probe are used per slide containing metaphase spreads or interphase nuclei (Jiang & Katz, 2002). Such analysis is normally performed on a single patient sample, thus the cost of a single analysis is extremely high, as 10 µl of probe cost 100 $. The development of a high throughput device for metaphase or interphase FISH analysis benefits from reduced probe volume per single sample, at the same time reducing the cost per diagnosis. Also, addressing the need for reduction in probe volume for single analysis can greatly increase the application of FISH in routine clinical diagnostics. Moreover, other standard cytogenetic analysis methods, such as karyotyping, also lack the automation. The introduction of automated microfluidic assays for cytogenetic analysis can offer more thorough and routine diagnosis that can be performed in the doctor's office at a lower cost and shorter time.

1.2 Microtechnologies in the cytogenetic field

Traditional cytogenetic analysis has evolved from karyotyping, through FISH techniques, Comparative Genomic Hybridisation (CGH), towards DNA microarrays. A few years ago, a routine chromosome test was carried out by culturing of the patients' blood sample followed by karyotyping, thorough banding analysis and validation by FISH. Nowadays, the first step performed in cytogenetics labs is often a CGH array, which is a genome wide DNA microarray that enables detection of deletions and duplications. It allows for assessment of the chromosome disorders by targeting multiple chromosome regions at once. At first the cytogenetic society was skeptical about their use; however their popularity has gradually increased over the years. Owing to that, microtechnologies gained trust in the cytogenetic scientific community and are now widely accepted.

Unlike microarrays, FISH can only detect few DNA regions in a single experiment, but introducing new microfabricated assays for interphase and metaphase FISH can greatly increase the use of such analysis. However, it should be noted that before these devices will reach cytogenetic laboratories all over the world they need to be tested for high quality and reproducibility of results.

In recent years the integration and automation of cytogenetic techniques has gained more attention. Most reports in this field focus on the development of an integrated microfluidic chip for interphase and metaphase FISH analysis. There are also some reports on the cell culture systems for suspension cells required for cytogenetic analysis. The recent developments in the field of microcytogenetics that address the need for automation, time and cost reduction in chromosome analysis are described here. Moreover, the commercially available machines and assays are also presented.

1.3 Cytogenetic analysis

A typical procedure for cytogenetic analysis is shown in **Figure 1**. The blood sample is collected from the patient and cultured for 3 days. The culturing step normally performed in

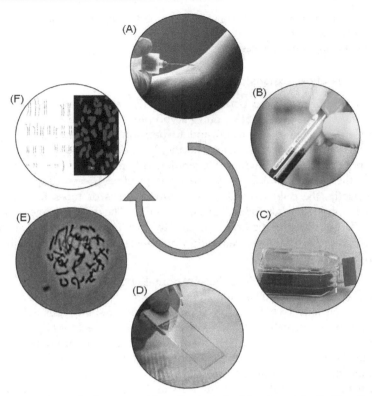

Fig. 1. The cytogenetic analysis starts with culturing of a blood sample, further fixation and splashing, followed by metaphase FISH or karyotyping.

culture flasks is prone to miniaturization and automation. The microdevices for culturing of suspension cells are presented in the following section. The next step requires harvesting of the chromosomes and chromosome spreads preparation on a glass slide. It is traditionally performed manually by dropping a cell suspension on a glass slide. Some machines and microdevices exist, which enable automation of the process. Finally, the analysis of chromosome glass slides is performed by either karyotyping or FISH. The conventional analysis requires the use of coplin jars for washing and incubation steps and usually uses high volumes of expensive reagents such as fluorescent probes. This traditional analysis protocol is far from automated but some examples of the microcytogenetics devices are available and described in this chapter.

2. Microchips for cytogenetic analysis

In this section we describe all available microdevices that enable performing a cytogenetic analysis. Firstly, the microfluidic bioreactors for culturing of suspension cells required for some protocols such as metaphase FISH or karyotyping. The main advantage of the presented cell culture microdevices is the ease of exchanging the liquids, i.e. from cell culture medium to hypotonic solution and finally fixative. All the laborious centrifugation steps can be omitted, which results in a higher yield of cells for the analysis. Further, the existing machines for metaphase chromosome spreads preparations are described. Lastly, the existing chips for performing miniaturized FISH are presented; with examples of FISH results obtained using these devices.

2.1 Sample preparation: Expansion, arrest and fixation of chromosomes

Cell culture is the primary step performed on patient samples containing lymphocytes, before they undergo metaphase FISH and karyotyping. The primary purpose of this cell culture is to ensure sample expansion and to perform sample preprocessing steps like hypotonic state inducement and arrest of the lymphocytes. Some of the cytogenetic laboratories perform the culture without separation of lymphocytes from red blood cells and plasma, but others prefer to culture lymphocytes purified by centrifugation with Ficoll-Paque®. Normally, the cells are cultured in culture flasks or tubes for 72 hrs in RPMI medium in a humidified incubator at 37 °C and 5 % CO_2. The disadvantage of such a culture method is the large volume of medium used and the fact that handling of suspension cells is tedious. Moreover, the traditional cell culture increases the risk of contamination due to manual sample handling.

To automate this process and reduce the contamination risk and volume of the medium a microfluidic device needs to be considered. Microfluidics-based cell culture devices due to small spatial dimensions have the promising prospect of providing cells with an *in vivo* like microenvironment in microchips. Moreover, for cells which are typically perfused actively via the vascular network, a perfusion based culture system provides a much better alternative to the standard static Petri dish-based cultures. By actively controlling the microenvironments surrounding the cells, we can exert greater control on the cell-cell interaction, the supply of nutrients to the cells and actively remove the biological waste (Kim et al., 2007).

Typically, shear stress applied to the cells is a major issue in case of microfluidic cell cultures and in case of suspension cell cultures the major challenge is to retain the cells in the system. A simple cell culture chip suitable for suspension cells was presented by Liu et al., (2008). They have fabricated a microfluidic device with minimum shear stress. The device was fabricated in two layers of PDMS bonded to a clean glass slide. The device consists of a main channel and side chambers for cell culturing. The medium supply and waste removal is achieved by convective and diffusive mass transport. The culturing of T-lymphocytes was demonstrated without cell losses due to shear stress. The main advantage of this design is that the cell medium perfusion can be started immediately after injection of cells, as the shear stress is too low to remove the cells. The main drawback of this device was the difficulty in extracting the cells for subsequent analysis post culture.

Recently, membrane based microfluidic bioreactors addressing all these issues and providing a simpler protocol for preparation of chromosome spreads were developed (Shah et al., 2011a; Svendsen et al., 2011). The proposed diffusion based microreactor (**Figure 2**) facilitates culturing of lymphocytes but also expansion, hypotonic treatment, and fixation of cells with the possibility to avoid several tedious centrifugation steps (Svendsen et al., 2011). Svendsen et al. developed a membrane based bioreactor for culturing a suspension of cells above the membrane with a microfluidic channel for media perfusion from the bottom. The cell culture in the bioreactor is performed for 72 hrs on lymphocytes purified from the

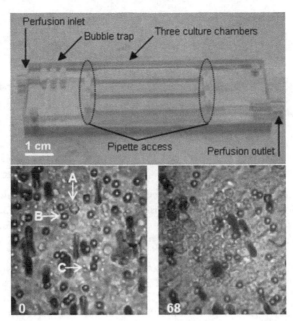

Fig. 2. The membrane based cell culture bioreactor for suspension cell culture. The two pictures below show a bright field image of the cells on the porous membrane at time 0 and after 68h of culturing. At the beginning the cells (A) are just slightly bigger than the pores (B). Platelets (C) can also be observed on the membrane (Reprinted with permission from Svendsen et al., 2011).

blood. The bioreactor was designed to provide pipette-based seeding and unloading of the cells. This ensures easy adaptability for the technicians who are familiar with pipette based traditional techniques. The inlet/outlet ports were sealed using a PCR tape during the cell culture. This is the only step, which requires manual operation of the device; all further steps are performed on a closed chip, which reduces contamination risk. The continuous perfusion of the medium ensures that all the necessary nutrients are delivered to the cells via diffusion from the membrane. It enables fast solution changes for expansion and cell fixation to obtain high quality metaphase spreads. Separation of the culture chamber by a membrane from the perfusion channel is also helpful to protect the cells from air bubbles formed in the flowing medium and allows the perfusion to be started even before the cells settle on the membrane. Svendsen et al. concluded that the cell growth inside the bioreactor was comparable to the control sample with the cells grown in a well-plate. However, the authors have not tested whether the culturing time can be reduced by means of this microfluidic bioreactor.

Shah et al. modified the media perfusion channel to ensure more thorough transport of nutrients across the membrane to the resting cells (Shah et al., 2011a). The pipette accesses were changed into microfluidic inlet and outlet ports connected to 3-port valves which allowed for easy removal of bubbles and change of medium (**Figure 3**). Shah et al. for the first time demonstrated that cell proliferation on the chip is better than in the control experiment on a Petri dish culture (**Figure 3B**).

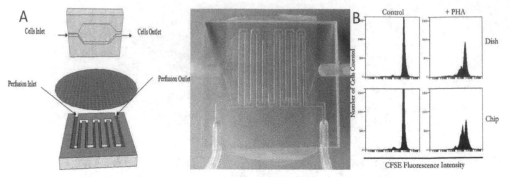

Fig. 3. A-The modified membrane based cell culture device with a perfusion channel that ensures better nutrients delivery to the cells. B-A graph showing proliferation of the CFSE stained cells in the device and on a Petri dish, control experiments without PHA stimulation (Reprinted with permission from Shah et al., 2011a).

One of the further steps required for cytogenetic analysis is harvesting of metaphase cells by hypotonic treatment followed by fixative addition. All these steps are traditionally performed by a series of centrifugation steps, which enables exchange of the solutions. The cultured cells are first centrifuged to remove the cell culture medium and to change the solution to hypotonic buffer. Such a treatment ensures cell swelling necessary for breaking of the membrane during preparation of chromosome spreads on slides. In order to make the membrane permeable the fixative, which is a mixture of methanol:acetic acid in a 3:1 ratio, is added to the pellet after centrifugation; this procedure is repeated up to 3 times. Harvesting of the metaphase chromosomes requires several laborious centrifugation steps, which have to be performed manually by a technician. A company called Transgenomic addressed this issue by introducing Hanabi-PII Metaphase Chromosome Harvester macromachine, which enables the steps to be performed automatically with more consistent results. Also, the presented microfluidic cell culture devices enable easy swelling and fixation of the cells. By simple change in the solution that is perfused below the membrane the hypotonic treatment and further fixation of the cells is performed on chip, in steps of 25 min. This is a simple and a very effective way of reducing the need for trained technicians and opens a possibility of performing a point-of-care analysis.

2.2 Chromosome spreads preparation

For reliable results cytogenetic analysis needs to be performed on a high quality sample. One of the very important steps in the cytogenetic analysis is the preparation of the high quality chromosome spreads on the glass slide. The technique, which is often used, is traditionally called 'splashing'. It is performed manually by skilled technicians and is greatly dependent on the environmental conditions such as temperature or humidity. Many cytogenetic laboratories have designated conditioned rooms to ensure non varying conditions for spreads preparation.

A traditional way of slides preparation varies from lab to lab. However, the splashing is commonly performed on glass slides that are kept, prior to the experiment, in a water container in a fridge to ensure proper wetting of the surface. A single use plastic Pasteur pipette is used to collect the cells in the fixative, which are further dropped on a tilted glass slide. The excessive liquid is drained on a tissue paper by placing the slide on the side. Afterwards, the slides are dried on a hot plate to facilitate evaporation of the fixative, which in turn enables spread formation.

There are a lot of factors that are presumably affecting the chromosome spreading process. In recent years, there has been considerable interest in fathoming the underlying process of chromosome spreading (Chattopadhyay et al., 1992; Gibas et al., 1987; Sasai et al., 1996). These explorations have also resulted in a number of devices aimed at automating the chromosome spreading protocol (Henegariu et al., 2001; Qu et al. 2008; Yamada et al., 2008). The most important environmental conditions for obtaining high quality spreads are temperature and humidity. These are often controlled by special humidity chambers with a temperature control. Another factor is a proper technique for slides ageing. The most common method is a simple air drying for 2 days, which helps the visualization of chromosome structures. Another technique involves dry-heat and chemical ageing, the latter being beneficial for FISH analysis, as it results in better signals and preserves chromosome architecture (Claussen et al., 2002, Henegariu et al., 2001; Rønne, 1989). Another factor suspected of affecting chromosome spreads quality is the dropping height, but Claussen et al. claim it is not essential. Last but not least, the presence of a thin water layer on the slide before dropping the cell suspension induces cell swelling resulting in better spreads.

Air drying is difficult to achieve in a completely closed microsystem, as the rate of fixative evaporation is crucial for good chromosome spreading. Based on this Vedarethinam et al. fabricated a splashing device, which contains two channels, one for ice cold water and one for cell suspension dropping (**Figure 4**). The injection of these solutions can be done manually, but is amenable for automation by use of syringe pumps. The system is designed to have a fixed dropping height and enables introduction of a thin water layer to the slide.

A ZenTech company produces the ZENDROPPER® macromachine, which is used for preparation of chromosome spreads with high reproducibility. Moreover, it enables high

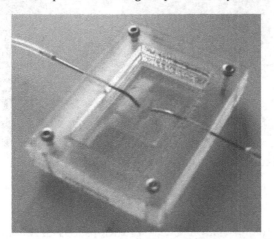

Fig. 4. A splashing microdevice for chromosome spreads preparation (Reprinted with permission from Vedarethinam et al., 2010)

throughput of 40 slides in 40 minutes preparation, which is essential for routine diagnostics laboratories. It allows for variations in the splashing height, but assures proper environmental conditions. However, the price and the size of this equipment prevent its common use in the laboratory, but anyone can benefit from the small sizes of the microdevices.

2.3 Integrated culturing, harvesting and splashing device

Recently, Shah et al. described a novel microfluidic FISHprep device for metaphase FISH slides preparation (**Figure 5**). The device combines the bioreactor for cell culturing (Shah et al., 2011a) with the splashing device for preparation of the chromosome spreads (Vedarethinam et al., 2010). This device consists of a diffusion based cell culture reactor separated from the splashing device by a clip valve. In this device, a 72 hr culture of T-lymphocytes with a CFSE staining was performed to determine the proliferation rate. Further, by means of addition of a clip valve which can be opened after the culture, the cultured and fixed lymphocytes can be splashed on the glass slide. The quality of spreads obtained from the FISHprep was comparable to traditionally obtained spreads. The integration of all the steps required for successful metaphase FISH slide preparation in one FISHprep offers a possibility for an automation of the molecular cytogenetics in future (Shah et al., 2011b).

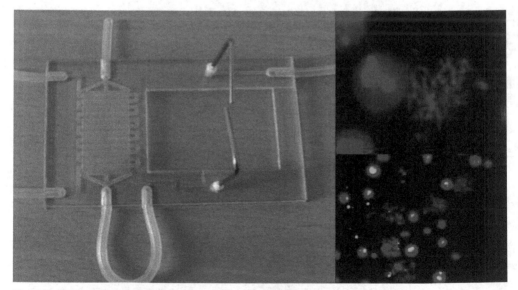

Fig. 5. Integrated cell culture chamber with a splashing device for preparation of the chromosome spreads for FISH analysis (Reprinted with permission from Shah et al., 2011b)

2.4 FISH on chip

In this subsection all commercially available microchips for FISH are introduced. Moreover, examples of the developed devices from the literature are also presented here. These microdevices address the need for probe volume reduction. Some of them miniaturized the interphase FISH protocol, but a few allow for metaphase chromosomes analysis. One of the interphase chip examples, tried to reduce the hybridization time, thus addressing the issue of long analysis time.

2.4.1 Cytocell multiprobe haematology

One of the very first miniaturized assays available on the market was provided by CytoCell. The devices called Chromoprobe Multiprobe Haematology are designed for FISH analysis of interphase and metaphase chromosomes from cultured peripheral blood cells or bone marrow samples. The Chromoprobe technology uses DNA FISH probes that are reversibly bound to the glass slide. During the experiment the probes dissolve in the solution of hybridization buffer and are readily hybridized to the chromosome preparations. The required denaturation of both chromosome preparations on the slide and probe in the solution occurs in the device itself during heating. This system allows multiple FISH probes to be hybridized to the same chromosome preparation, which enables rapid screening for multiple abnormalities in a single experiment. Such a solution makes FISH analysis easier and quicker, however as the slide is divided into small squares with different immobilized probes it requires troublesome analysis and counting of the results.

2.4.2 BioCellChip

In 2007, Lee et al., presented one of the first examples of miniaturized devices for performing FISH analysis (Lee et al. 2007). This device in **Figure 6** is suitable for interphase FISH, allowing high throughput analysis on a cell array. The authors have shown the possibility to array cells on the glass slide by spotting small amounts of samples onto a supporting PDMS matrix. The chip consists of a patterned glass slide with bonded 1 mm perforated PDMS layer. The glass slide and PDMS lid form an array of 96 cavities of 1.5 mm in diameter for cell spotting. The patterning of a glass slide with 96 squares and numbers for indexing of wells was achieved by a photolithography and thin-film metal deposition process. The PDMS layer was only used for a controlled cell sampling on the glass slide, mainly to stop the cell samples from spreading all over the glass slide. After air drying, the conventional FISH protocol was performed on a glass slide, without the supporting PDMS layer. The main advantage of this device is that for 96 different specimens only 10 µl of the probe can be used. Such a microfabricated bio-cell chip with a PDMS layer is useful in mass screening of microdeletions or aneuploidy with the same probe used for detection of chromosome abnormalities. The authors have mentioned the probe volume reduction used to analyze 96 specimens in one experiment, but other major issues such as manual intervention and time consumption were not stated. Moreover, it is discussed that PDMS could be replaced by cement or paper stickers, to create the spotting chambers and physically separate them.

2.4.3 Interphase FISH on chip

The first demonstration of FISH on chip was showed by Sieben, et al. (2007). They have adapted a conventional interphase FISH protocol to a miniaturized version (**Figure 7**). One of the few differences was immobilization of cells, which was achieved by temperature treatment. Moreover, the authors explored the methods of hybridization time reduction by recirculation of the probe by means of on-chip peristaltic pumps and electro-kinetic transport by external electrodes inserted in the end wells. It was shown that electro-kinetic

Fig. 6. BioCellChip enabling high throughput analysis of several cell samples (Reprinted with permission from Lee et al., 2007)

probe circulation, which is easy to integrate into the design, performed better than the recirculation. The recirculation method involved a use of on chip valves and pumps, which require complex microfabrication method. Nonetheless, by performing FISH on-chip at microfluidic volumes, they were able to achieve a tenfold reduction in DNA probe consumption per test, with corresponding reduction of a single experiment cost. In 2008,

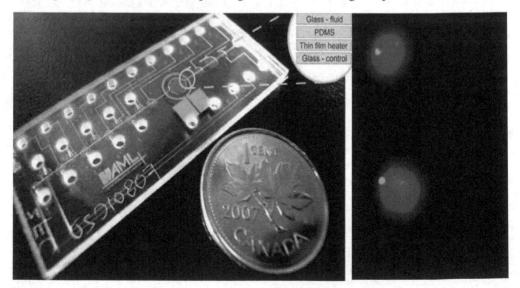

Fig. 7. Interphase FISH on CHIP with integrated mixing elements to enhance the hybridization. The FISH results showing a male sample with one X and Y chromosomes (Reprinted from Sieben et al., 2008 with permission of The Royal Society of Chemistry)

Sieben, et al., published another paper in which all previously performed steps with microfluidic FISH were integrated onto one chip and automated using valves and a computer controlled peristaltic pump (Sieben et al., 2008). Whereas conventional FISH takes nearly an hour of on-and-off attention from a skilled technician, using their time inefficiently, automated microfluidic FISH reduced human intervention to only involve reagent preparation, and results analysis with fluorescence microscope. However, the chip design is very complex and technicians would need additional training to be able to use the device.

2.4.4 Miniaturized interphase FISH

The most recent example of an interphase FISH on chip was presented by Zanardi et al., (2010). They presented a miniaturization of the interphase FISH protocol through microfluidic methods with minimized shear stress exerted on cells (living or fixed) during sample loading. The authors focused on the use of biomaterials and coating to promote cell adhesion, mainly nanostructured TiO_2 coating that triggers a rapid and efficient cell immobilization.

The microfluidic device consists of a PDMS pad with a straight 0.15 µl microchannel with an inlet and outlet that are accessible with an automatic pipette tip. The PDMS pad was manually bonded to a glass slide previously coated with 50 nm of TiO_2 and treated with oxygen plasma to increase hydrophilicity. The coating was achieved by means of cleanroom cluster beam technology, which greatly increases the device fabrication costs. Sample loading into the device was achieved by simply pipetting 1.5 µl of cell sample into the inlet hole and left to enter the channel by capillary force. The entire FISH protocol was performed in the microfluidic device, by passing all the necessary solutions through the channel. After injection of the probe, the inlet and outlet holes were sealed with a drop of mineral oil for overnight probe hybridization to prevent the evaporation. Post hybridization washing steps were performed in a conventional ways on the TiO_2 coated slide without a PDMS pad. The device was fully validated using various clinical samples to analyze chromosome abnormalities. The device performance was compared to that of the conventional method. The miniaturized version of FISH provided accurate, high quality, reproducible results as compared to standard FISH protocol. The main advantage of this microchip is its ease of use and reduction in the probe volume used (10- to 30-fold less probe compared to the standard protocols). Moreover, due to the confinement of cell sample in a relatively small area, the image analysis can be performed rapidly in less than 8 minutes, which is 5 times faster than image analysis after conventional FISH. The device gained attention in the cytogenetic field and is now available commercially as the product microFIND research® from the TETHIS S.p.A. company. It enhances cell retention and a uniform distribution of the cells inside the channel, reducing the reagents volume up to 20 times.

2.4.5 Metaphase FISH on chip

The first implementation of a chip for metaphase FISH analysis was presented by Vedarethinam et al., (2010). Metaphase spreads formation is based on the hydrophilicity of the substrate, a proper rate of evaporation, temperature gradient, humidity and controlled splashing angle, which is difficult to achieve in a closed chip (Deng et al., 2003; Henegariu et al., 2001). The authors have presented a novel splashing device with an open chamber, which is used for dropping the fixed cell sample and ice cold water on the glass slide. In this way the evaporation is easily controlled and the chromosome spreads achieved by the splashing device are comparable to those achieved by the conventional method. By placing a double adhesive tape stencil on the glass slide the confinement of the metaphase spreads was achieved. The adhesive tape serves as a bonding support for mounting a PDMS lid with two microchannels and a reaction chamber (**Figure 8**). Such a rapid and easy protocol for the microFISH device allows for a quick transformation of a simple glass slide into a metaphase FISH on chip assay. All steps in a conventional metaphase FISH protocol, such as washing, dehydration, probe injection, were performed on the chip. The authors focused mainly on the miniaturization of a conventional protocol, without introducing any innovation to the existing method. The advantage of the microFISH device over traditional method is the 2-fold reduction in the probe volume used, which reduces the cost of the analysis. The presented device is a good solution for integration into existing work routines in cytogenetic labs. Moreover, it is amenable for automation by preloading of the reagents into tubings. The reagents preloading into tubings was tested for a further reduction of a probe volume showing some promising results (unpublished data), which can allow for a widespread use of the microFISH device.

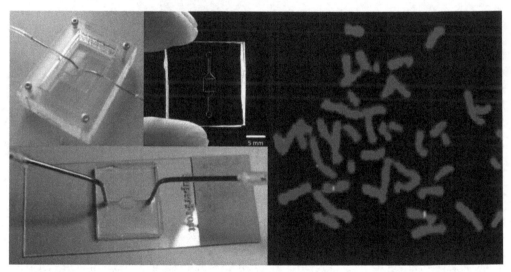

Fig. 8. Metaphase FISH on chip with an open splashing device to enable spread formation. The PDMS lid consists of two channels with a chamber for performing FISH. On the right FISH results with X centromeric probe inside the chip are shown (Reprinted with permission from Vedarethinam et al., 2010).

2.5 Chromosome sorting on chip

The cytogenetic analysis often reveals the chromosome changes present in the sample. Very often the resolution of the techniques is not sufficient to determine the small changes. Cytogenetists often need to obtain the derivative chromosomes carrying the abnormalities for further testing by e.g. next generation sequencing or for recombinant DNA libraries construction. One of the most well known techniques is the chromosome sorting by flow cytometer. However, many disadvantages such as need for high sample volume, lack of precision and poor reproducibility, have hindered its common use. Moreover, the flow sorters are expensive and require skilled technicians to operate the machines.

Recently, an attempt to replace the traditional chromosomes flow sorter with a microdevice has been reported by Inoue et al., (2008). The authors presented two methods suitable for manipulating the chromosomes – by electric and magnetic fields. The voltage modulation was suggested to replace the conventional gel electrophoresis for chromosomes sorting. The mobility of the chromosomes depends on their size when placed in a time-dependent electric field, with the smallest chromosomes moving faster. As the chromosomes size distribution is very large, the authors classified them into three size based groups. They proved that during 30 minutes experiment in the device the chromosomes have been sorted into these 3 groups, however, high sample density resulted in clogging.

The other presented method requires use of a magnetic field. The authors fabricated a device for continuous flow sorting of chromosomes, with high throughput and high operation speed. The chromosomes labelled with superparamagnetic beads are dragged by a magnetic field against the laminar flow stream and are collected at separate outlets. The force is proportional to the amount of magnetic beads attached to the chromosomes, which in turn depends on the chromosome size. The deflection of larger chromosomes is bigger than that for small chromosomes. Further studies are conducted to improve the chromosome sorting accuracy and throughput by means of the microfluidic device.

3. Future prospects

While cytogenetic analysis continues to be cumbersome, manual and expensive, the recent advances in the microtechnologies enabling cytogenetics offer a new direction and hope for optimizing the protocols and making them simpler, cheaper and available on a wider scale. The successful demonstrations of devices for the various steps involved in the cytogenetic procedures need to now be tested and validated clinically before they can be implemented on a routine basis. The microfluidic bioreactors have proved superior for application in suspension cell cultures, which opens up possibilities for utilizing them for culture of amniocytes or chronic villi, which will speed up the prenatal diagnosis protocols. The device for metaphase FISH offers good prospects for automating the sequential FISH sample injection protocol leading to automated FISH. In the distant future, greater benefits of using microtechnologies for cytogenetics will be reaped when these assays will be automated and multiplexed; and be able to process multiple samples simultaneously.

4. Conclusion

Cytogenetic analysis performed nowadays is a laborious, technically demanding and long process, starting from the culturing of lymphocytes from patient sample through preparation of chromosome spreads to FISH or karyotyping analysis. All these steps are performed manually by skilled technicians; however, to perform the test routinely in the doctor's office there is a need for automation and simplification of the protocol. One of the main drawbacks of FISH is its complexity and price of the probes, which hinders the usage of this technique. Owing to that, DNA based techniques are gaining more popularity, and are nowadays the first analysis step performed routinely in cytogenetic laboratories. FISH is now just a supplementary technique used for validation of the results. Although, some

Microtechnology in Cytogenetic Analysis		
Sample preparation	Liu et al., 2008	*Minimal shear stress cell culture device* Cell culture chambers on a side of the main channel Convective and diffusive nutrient supply Shear stress minimized Difficult cell extraction
	Svendsen et al., 2010	*Membrane cell culturing bioreactor* Adherent and Non-Adherent cells Pipette access holes for cell loading/unloading Cell culture separated from the medium perfusion channel Culture time – 72 hrs Continuous diffusion based perfusion Easy solution changes
	Shah et al., 2011a&b	*Microfluidic Bioreactor for cell culture* *In vivo* like environment Cell culture separated from the medium perfusion channel Culture time – 72 hrs Continuous diffusion based perfusion Better cell growth Easy operation Easy cell retrival Easy solution changes
Glass Slide Preparation	Vedarethinam et al., 2010, Shah et al., 2011b	*Splashing device* Cold water and cell inlets Reliable metaphase spreads Suitable for automation Easy operation
FISH	CytoCell	*Multiprobe Haematology* Interphase/metaphase FISH Sample spotting on a glass slide Probes reversibly bound to the glass slide Conventional washing steps
	Lee et al., 2008	*BioCellChip* Interphase/metaphase FISH 96 cell samples spotted using PDMS stencil Conventional washing steps Conventional hybridization protocol Probe volume - 10 µl Probe volume reduction 10 ul for 96 samples High throughput
	Sieben et al., 2007&2008	*Interphase FISH on chip* Interphase FISH Probe volume reduction 10- to 20-fold Automated protocol Complicated design Expensive fabrication

Microtechnology in Cytogenetic Analysis		
	Zanardi et al., 2010	*Miniaturized interphase FISH* Interphase FISH Cell immobilization on TiO_2 coated glass slide PDMS channel lid Probe volume – 0.3 μl Probe volume reduction – 10- to 30- fold Reduced analysis time Suitable for automation and high throughput
	Vedarethinam et al., 2010	*Metaphase FISH on chip* Interphase/metaphase FISH Glass slide preparation using splashing device PDMS channel lid Probe volume – 5 μl Probe volume reduction – 2 fold Suitable for automation and high-throughput Inexpensieve

Table 1. A summary of available microdevices for cytogenetic analysis

machines for FISH protocol automation exist, their cost and size prevent their usage in non-laboratory settings. Table 1 summarizes the advantages and disadvantages of the microdevices available for cytogenetic analysis described in this chapter.

These microdevices and assays can be easily incorporated in the routine cytogenetic lab environment, but can also be available for point of care diagnostics. The cost of such assays is greatly reduced, due to smaller volumes of probe used. Finally, each step of the protocol can be transformed to a microdevice format allowing a total analysis of chromosomes to be done on a microplatform. Such a miniaturization of a protocol enables high-throughput analysis of several samples at the same time and could drastically speed up the process to provide fast results. Moreover, the manipulation of the probe across the chromosome spreads inside microfluidic devices could reduce the hybridization time down to minutes.

5. References

Chattopadhyay, S.; Marinduque, B. and Gilbert, F. (1992). Ovarian-Cancer—Protocol for the Preparation of Banded Metaphases from Tumor-Tissue. *Cancer Genetics and Cytogenetics*, Vol. 59, pp. 210-212.

Claussen, U.; Michel, S.; Muhlig, P.; Westerman, M.; Grummt, U.-W.; Kromeyer-Hasuchild, K. and Liehr, T. (2002) Demystifying chromosome preparation and the implications for the concept of chromosome condensation during mitosis, *Cytogenetic and Genome Research*, Vol. 98, pp. 136-146

Deng, W.; Tsao, S.W.; Lucas, J.N.; Leung, C.S. and Cheung, A.L.M. (2003). A new method for improving metaphase chromosome spreading. *Cytometry Part A*, Vol. 51A, No. 1, pp. 46-51.

Gibas, L.M.; Grujic, S.; Barr, M.A. and Jackson, L.G. (1987). A Simple Technique for Obtaining High-Quality Chromosome Preparations from Chorionic Villus Samples Using FDU Synchronization. *Prenatal Diagnosis*, Vol. 7, pp. 323-327.

Henegariu, O.; Heerema, N.A.; Wright, L.L.; Bray-Ward, P.; Ward, D.C. and Vance, G.H. (2001). Improvements in cytogenetic slide preparation: Controlled chromosome spreading, chemical aging and gradual denaturing. *Cytometry*, Vol. 43, No. 2, pp. 101-109.

Inoue, T.; Fujita, Y.; Uchiyama, S.; Doi, T.; Fukui, K. and Yokoyama, H. (2008). On-Chip Chromosome Sorter Using Electric and Magnetic Fields, In: *Chromosome Nanoscience and Technology*, Fukui, K. and Ushiki, T., pp. 43-52, Taylor & Francis Group, ISBN 978-1-4200-4491-1

Jiang, F. & Katz, R.L. (2002). Use of interphase fluorescence in situ hybridization as a powerful diagnostic tool in cytology. *Diagnostic Molecular Pathology*, Vol. 11, pp. 47-57.

Kim, L.; Toh, Y.-C.; Voldman, J. and Yu, H. (2007). A practical guide to microfluidic perfusion culture of adherent mammalian cells. *Lab on a Chip*, Vol. 7, pp. 681-694

Lee, D.S.; Lee, J.H.; Min, H.C.; Kim, T.Y.; Oh, B.R.; Kim, H.Y.; Lee, J.Y.; Lee, K.C.; Chun, H.G. and Kim, H.C. (2007). Application of high throughput cell array technology to FISH: Investigation of the role of deletion of p16 gene in leukemias. *Journal of Biotechnology*, Vol. 127, pp. 355-360.

Liu, K.; Pitchimani, R.; Dang, D.; Bayer, K.; Harrington, T. and Pappas, D. (2008). Cell Culture Chip Using Low-Shear Mass Transport. *Langmuir*, Vol. 24, pp. 5955-5960

Qu, Y.Y.; Xing, L.Y.; Hughes, E.D. and Saunders, T.L. (2008). Chromosome Dropper Tool: Effect of Slide Angles on Chromosome Spread Quality for Murine Embryonic Stem Cells. *Journal of Histotechnology*, Vol. 31, pp. 75-79.

Rønne M. (1989). Chromosome Preparation and High Resolution Banding Techniques. A Review. *Journal of Dairy Science*, Vol. 72, No. 5, pp. 1363-1377

Sasai, K.; Tanaka, R.; Kawamura, M.; Honjo, K.; Matsunaga, N.; Nakada, T.; Homma, K.; Fujimura, H. (1996). Chromosome Spreading Techniques for Primary Gastrointestinal Tumors. *Journal of Gastroenterology*, Vol. 31, pp. 505-511

Sieben, V.J.; Debes -Marun, C.S; Pilarski, P.M.; Kaigala, G.V.; Pilarski, L.M. and Backhouse, C.J. (2007). FISH and chips: chromosomal analysis on microfluidic platforms. *IET Nanobiotechnology*. Vol. 1, pp. 27-35.

Sieben, V.J.; Debes-Marun, C.S.; Pilarski, L.M. and Backhouse, C.J. (2008). An integrated microfluidic chip for chromosome enumeration using fluorescence in situ hybridization. *Lab on a Chip*, Vol. 8, pp. 2151-2156.

Shah, P.; Vedarethinam, I.; Kwasny, D.; Andresen, L.; Dimaki, M.; Skov, S. and Svendsen W.E. (2011a). Microfluidic bioreactors for culture of non-adherent cell. *Sensors and Actuators B:Chemical*, Vol. 156, No 2, pp. 1002-1008.

Shah, P.; Vedarethinam, I.; Kwasny, D.; Andresen, L.; Skov, S.; Silahtaroglu, A.; Tumer, Z.; Dimaki, M. and Svendsen W.E. (2011b). FISHprep: A novel Integrated Device for Metaphase FISH Sample Preparation. *Micromachines*, Vol 2, No. 2, pp. 116-128.

Svendsen, W.E.; Castillo-Leon, J.; Lange, J.M.; Sasso, L.; Olsen, M.H.; Abaddi, M.; Andresen, L.; Levinsen, S.; Shah, P.; Vedarethinam, I. and Dimaki M. (2011). Micro and nano-platforms for biological cell analysis. *Sensors and Actuators A:Physical*, Article in press, ISSN 0924-4247, doi:10.1016/j.sna.2011.02.027

Vedarethinam, I.; Shah, P.; Dimaki, M.; Tumer, Z.; Tommerup, N. and Svendsen, W.E. (2010). Metaphase FISH on a Chip: Miniaturized Microfluidic Device for Fluorescence in situ Hybridization. *Sensors*, Vol. 10, pp. 9831-9846.

Yamada, K.; Kakinuma, K.; Tateya, H. and Miyasaka, C. (2008). Development of an Instrument for Chromosome Slide Preparation. *Journal of Radiation Research*, Vol. 33, pp. 242-249.

Zanardi, A.; Bandiera, D.; Bertolini, F.; Corsini, C.A.; Gregato, G.; Milani, P.; Barborini, E. and Carbone R.. (2010). Miniaturized FISH for screening of onco-hematological malignancies. *BioTechniques*, Vol.49, pp. 497-504.

5

Array CGH in Fetal Medicine Diagnosis

Ricardo Barini, Isabela Nelly Machado and Juliana Karina R. Heinrich
Fetal Medicine Program, Cytogenetics Core, Women´s Hospital,
State University of Campinas, UNICAMP, Campinas, SP
Brazil

1. Introduction

About 2% to 5% of live births have at least one identifiable congenital anomaly at birth (Kalter and Warkany, 1983), ranging from mild to severe abnormalities that compromise the survival. Congenital malformations have been showing increasing importance as a cause of suffering and harm to health of the population, accounting for a large percentage of perinatal morbidity and mortality (De Galan-Roosen et al., 1998, Rosano et al., 2000, Cornel, 2000).

The introduction of ultrasonography in prenatal care allowed the identification of the congenital malformations even in the intrauterine environment. The identification of malformations occurs directly by the detection of morphological changes or indirectly through signs such as fetal growth retardation and changes in amniotic fluid volume. Technological advances in ultrasound, growing experience for specialized services and easier access to ultrasound services have contributed to a significant increase in the detection of fetuses with congenital malformations in low risk populations, becoming a routine part of prenatal care. The ultrasound to identify fetuses with chromosomal abnormalities showed a method with high negative predictive value, which is to say that in the absence of defects detected, the possibility that the fetus has a chromosomal abnormality is low (Nicolaides et al., 1992, Boue et al., 1988, Wladimiroff et al., 1988, Gonen et al., 1995). Adding to this evidence to the fact that the majority of affected fetuses are generated by young women with no identified risk factor (Nicolaides et al., 1992, Zeitune et al., 1991), ultrasonography with emphasis on the pursuit of fetal malformations that are related to chromosomal abnormalities is recommended in all pregnant women in more than one occasion during pregnancy.

The genetic causes, alone or in conjunction with environmental causes, are involved in at least one third of congenital malformations. Among the genetic causes, numerical or structural chromosomal abnormalities are present in about 0.9% of unselected group of newborn (Jacobs et al., 1992) and in more than 10% of stillbirths (Jackson, 2002). The overall frequency of cytogenetic abnormalities in malformed fetuses is approximately 10% to 15% (Nelson and Holmes, 1989). Of these, about 80% are trisomies of chromosomes 13, 18 or 21. The rest involves numerical changes in the sex chromosomes and structural chromosomal rearrangements (Rickman et al., 2006). It is justifiable, therefore, the conduct to indicate the chromosomal analysis of all stillbirths and neonatal deaths, with or without dysmorphic phenotypes, and all fetuses with malformations, especially when more than one anomaly is present or when there is family history of congenital malformations.

Since the description of the number of human chromosomes 50 years ago, there is a scientific effort to define the association between chromosomal abnormalities, genetic diseases and congenital malformations. The development of modern molecular techniques has allowed the analysis of the human genome in high resolution, especially useful in identifying new genomic changes, not previously identified by routine chromosome analysis.

Faced with a malformed fetus that has normal chromosomal study through G-banded karyotype, one must be very careful and avoid any rash attempts of syndromic diagnosis. Even after birth, with a complete physical exam and various laboratory tests, diagnostic centers specializing in dysmorphic syndrome cannot establish definitive diagnosis in about half of the patients (RE, 1993). For this group, molecular techniques involving genomic level studies allow the identification of new chromosomal microarray which might be responsible for abnormal phenotype, contributing to the molecular characterization and establishment of a more accurate diagnosis, a most appropriate perinatal approach and a more detailed genetic counseling.

2. Prenatal diagnosis of chromosomal abnormalities

Microscopic analysis of chromosomes has been the gold standard for diagnosis of chromosomal abnormalities since the development of G-banding technique in the late 60's (Caspersson et al., 1968). In a Fetal Medicine Unit, it is common to obtain a karyotype from umbilical cord blood, amniotic fluid and chorionic villi, depending on gestational age. However, theoretically, it is possible to obtain the karyotype from any tissue that has preserved its viability and can be subjected to cell culture to obtain metaphase (Cabral et al., 2001). The time to obtain the results vary depending on the material studied the viability of cultured cells, the technical laboratory, the reagents and sample preparation, among other factors. On average, this period is 7 to 15 days for the culture of chorionic villi or amniotic fluid and from 3 to 7 days for fetal blood (Nussbaum et al., 2008). The direct preparation of chorionic villi, without cell culture, allows us to obtain results within 24 hours, but the low resolution of the chromosome limits the application, making it impossible sometimes to exclude structural abnormalities.

Although very reliable and still considered the gold standard for the investigation of abnormalities related to chromosomes, the conventional cytogenetic analysis has some limitations, such as failure of cell culture and the need for experienced professional to read and reliably interpret results. In prenatal diagnosis, other important limitations are related to the low quality of chromosome preparations, obtained in a significant portion of the time, and which prevents the detection of structural anomalies and chromosomal microarray smaller than 5Mb to 10Mb. In addition, the need for long-term cell culture ultimately significantly impact the relatively long time for the release of results.

In recent decades, clinical cytogenetics has seen extraordinary advances in molecular biology techniques. The confluence of molecular and cytogenetic approaches - molecular cytogenetics - revolutionized the possibilities and diagnostic accuracy. The sequencing of the human genome has contributed to this effect, allowing a higher tracking resolution for chromosomal abnormalities. These genomic changes are about 15% of all mutations that involve single-gene diseases in humans (Vissers et al., 2005). The limitations of conventional banding resolution can be largely overcome by new molecular techniques, with obvious clinical applications in the diagnosis of microdeletion, subtelomeric rearrangements, marker chromosomes and derivatives, and gene rearrangements (Carpenter, 2001).

The most widespread molecular technique in prenatal diagnosis is fluorescence in situ hybridization (FISH), by assessing the presence or absence of a specific DNA sequence or chromosomal region. Therefore, the FISH technique is locus-specific and requires knowing the DNA sequence of interest to the appropriate choice of the probe to be used. It is used for early detection of trisomy in uncultured cells, simultaneously analyzing five probes for chromosomes frequently involved in aneuploidy (13, 18, 21, X and Y) and to confirm microduplications or microdeletions syndromes. The advantage over conventional techniques is the short time to obtain the results; it is possible in just 2 hours. Several other techniques have emerged based on the original method of FISH. Using the techniques of multicolor FISH (M-FISH) and spectral karyotyping (Spectral karyotyping - SKY), each chromosome takes a "signature" spectrum, enabling the identification of complex rearrangements involving more than two chromosomes, as well as the source and content of marker chromosomes (Haddad et al., 1998, Fan et al., 2000), with a resolution of 1Mb to 5Mb for the M-FISH (Speicher et al., 1996) and 1Mb to 2 Mb for SKY (Schröck et al., 1997).

The conventional cytogenetic techniques approach associated with the study of molecular dysmorphology have allowed a greater correlation between genotype and phenotype, with the diagnosis of an increasing number of "microarray syndromes" or "genomic instability". The term "genomic instability" has been widely used to describe a phenomenon that results in the accumulation of multiple changes that lead to conversion of a genome of a normal cell to an unstable genome (Smith et al., 2003). These unbalanced chromosomal rearrangements account for 1% to 2% of the abnormalities in prenatal samples, and can lead to severe phenotypic consequences (Ryall et al., 2001). The major limitation of molecular techniques described so far is that these techniques do not detect these "genomic instability".

3. Comparative genomic hybridization (CGH)

The development of microarray technology in the early 1990's at Stanford University (Schena et al., 1995) drew heavily from six major disciplines: Biology, Chemistry, Physics, Engineering, Mathematics and Computer Science. Besides no other technology has ever involved so much technological complexity, combining expertise from so many different disciplines, it provided a quantitative and systematic view of a biological system.

This revolutionary new science uses microscopic glass arrays (microarrays) for quantitative analysis of genes (or part of them) and gene products. It has its root in advances made between the discovery of DNA in 1950's and the Human Genome Project in 1990's.

The first hybridization experiments on glass were performed in the late 1980's and early 1990's (Khrapko et al., 1989, Fodor et al., 1991, Maskos and Southern, 1992, Lamture et al., 1994, Guo et al., 1994). These early experiments established the feasibility of glass-based hybridization, combinatorial oligonucleotide synthesis, linker and face chemistry; contact printing based on capillary action and early detection technologies. Many of these principles were used to develop the first microarray assays (Schena et al., 1995)

The comparative genomic hybridization (CGH) was developed as a method of comprehensive genome scan, in an attempt to identify imbalances in the number of copies of DNA, ie genomic instability. Developed in 1992, the CGH technique is the competitive hybridization of test DNA and normal DNA, marked with different fluorochromes (Kallioniemi et al., 1992). The original technique used normal metaphases in fixed blade and is known as metaphasic, chromosomal or conventional CGH.

The metaphasic CGH technique has its main applicability in the field of cancer genetics (Albertson and Pinkel, 2003, Weiss et al., 2003a, Weiss et al., 2003b). Other clinical applications as in dysmorphology, mental disorders and learning were also tested, showing an increase in the diagnosis of deletions or duplications not identified by G-banding (Kirchhoff et al., 2001, Ness et al., 2002). For prenatal diagnosis, the technique has been validated retrospectively studying fetuses known to be carriers or partial aneuploidy (Bryndorf et al., 1995, Yu et al., 1997, Lapierre et al., 1998, Thein et al., 2001), being equivalent to high-resolution cytogenetic techniques, but with the advantage of not requiring cell culture. A study demonstrated its utility as a complementary tool to conventional karyotype in a group of fetuses with abdominal wall defects (Heinrich et al., 2007). However, the delineation of specific regions and bands in conventional CGH technique is limited by the resolution of metaphase chromosomes and is estimated to be from 3Mb to 10Mb, and it is the most limiting (Kallioniemi et al., 1992, Bryndorf et al., 1995, Kirchhoff et al., 1999).

Keeping the same principle of comparing DNA test (sample) and normal DNA (reference) and coupling to the microarray technology (microarrays or arrays) (Solinas-Toldo et al., 1997, Pinkel et al., 1998), a new technique of comparative genomic analysis has been developed and it is now based in microarrays, known as array CGH. Basically, the technique uses multiple pre-selected fragments of DNA attached, so a locus-specific glass surface (microscope slide). The attached DNA can be cloned DNA fragments from bacteria (BAC - Bacterial Artificial Chromosomes) or P1 (PAC - P1 Artificial Chromosomes) (Telenius et al., 1992), cDNA (Pollack et al., 1999), synthetic oligonucleotides (Lucito et al., 2003) or PCR fragments (Mantripragada et al., 2004).

The type of DNA used and the amount of the surveyed areas vary between different protocols or platforms available. The choice of regions included in the survey defines their classification into two types: array CGH representative of the entire genome and targeted to specific regions, usually involved in chromosome arrangements already described. The resolution obtained by different types of array CGH depends on the number and size of clones studied and the distance between consecutive clones. Arrays constructed from BAC clones have the resolution of 50kb to 150kb (Ishkanian et al., 2004, Coe et al., 2009) and oligonucleotide arrays can come in the 25pb 85pb resolution (Shaikh, 2007).

Technically, identical amounts of DNA test and reference DNA are labeled with fluorochromes (cyanine), and after being co-hybridized, the difference of the intensities emitted by the cyanine for each area surveyed is measured. When the fluorescence intensity is lower in the sample tested against the reference sample, we infer that there is a loss in the respective genomic region ("deletion" or "loss") and in the opposite situation; it is inferred genomic gain ("double "or" gain).

The array CGH technique offers important advantages over other methods of cytogenetic diagnosis. First, because it is not necessary to obtain metaphases and allow the analysis of

DNA extracted from different tissues, even tissues embedded in paraffin. Other advantages include the ability to search thousands of chromosomal regions in a single analysis in a short period of time and with high resolution, overcoming the limitations of conventional karyotyping and metaphase CGH. Other advantages of this molecular method is that it does not require prior knowledge of the genomic region involved and the ability to study cases where only DNA is available and no chromosomes can be obtained. The method was reproducible in a clinical standpoint, with reliable results within 48 hours.

Its inherent limitation, based on the principles of the technique, lies in not being able to identify chromosomal abnormalities that do not lead to change in the total number of copies of a segment of DNA within the genome, such as balanced translocations and inversions. Also, the normalization of the doubled fluorescence intensity generated in euploidy can difficult their identification. The technique also finds limited use in cases of mosaicism. This last limitation has been overcome with increased experience in interpreting the results, and is believed to be possible to detect chromosomal rearrangements present in at least 20% (Vermeesch et al., 2005) or 50% of the cells (Shaw and Lupski, 2004). There is recent evidence that the technique of array CGH is able to diagnose cases of mosaicism that were not detected by conventional karyotyping (Ballif et al., 2006, Cheung et al., 2007, Shinawi et al., 2008, Wood et al., 2008).

Microarray-based CGH is a powerful method to detect and analyze genomic imbalances that are well below the level of detection on high resolution banded karyotype analysis, providing a better opportunity for genotype/phenotype correlations in other similarly affected individuals. The clinical application of array CGH as a diagnostic tool in patients with mental disorders and learning associated with dimorphic features has been extensively studied (Bejjani et al., 2005, Cheung et al., 2005 Jul-Aug, Shaffer et al., 2006, Lu et al., 2007, Stankiewicz and Beaudet, 2007), and has proven useful and reproducible for diagnosis and molecular characterization of this group (Vissers et al., 2003, Shaw-Smith et al., 2004, Gribble et al., 2005, Schoumans et al., 2005, Rosenberg et al., 2006, Menten et al., 2006, Aradhya et al., 2007, Pickering et al., 2008, Brunetti-Pierri et al., 2008). These studies have shown that array CGH technique can increase the rate of diagnosis in relation to conventional cytogenetic study in 20% to 30%, culminating in its inclusion in the flowcharts of genetic research of this group of patients (Sharkey et al., 2005). Microarray-based genomic copy-number analysis is now a commonly ordered clinical genetic test for patients with diagnoses including unexplained developmental delay/intellectual disability, autism spectrum disorders, and multiple congenital anomalies. It is offered under various names, such as "chromosomal microarray" (CMA) and "molecular karyotyping".

The increase in coverage of genomic regions in the array CGH platforms representative of the entire genome is responsible for an additional fee of 5% in the detection of chromosomal microarray compared to array CGH platforms targeted to specific regions (target arrays) (Baldwin et al., 2008). However, while the high-resolution analysis can identify pathological abnormalities, they incur the problem of identification of genomic gains and losses in regions with unknown clinical significance. So as you increase the resolution of the platform, simultaneously increases the degree of uncertainty. Here comes the delicate balance between resolution of the method, the potential for clinical diagnosis and the ability to interpret the results. This is the key issue to determine the use of CGH microarray for diagnostic use in clinical practice.

The CNV may be difficult to evaluate and interpret. A variety of methodologies used in the generation of information that form the research databases available electronically CNV makes it difficult to discern the exact extent of each CNV likely found. This lack of methodological uniformity may confuse the correct interpretation of finding abnormal chromosome present in a phenotypically abnormal individual, but overlapping all or part of a region with CVN described. Another problem is the finding of a gain in a region where it is described as a deletion CNV and vice versa. In these situations, we can not infer that the rearrangement is also a benign finding.

The identification of CNVs may also vary according to the CGH array platform used. A recent study showed that the number of CNVs found using the technique of array CGH containing BAC clones was higher when compared with those found when applied oligonucleotide array CGH (Aradhya et al., 2007). This discrepancy can be explained by the fact that the clones generated from bacteria (BAC) is that the larger fragments oligonucleotide probes, with the consequent overlapping regions. In addition, BAC clones can encompass regions of repetitive DNA, known to be associated with variations in the number of copies (Sharp et al., 2005).

However, at this point in time, genome-wide arrays will detect many copy number variants of unknown clinical significance. Growing clinical experience with genome-wide arrays and the development of copy number variants databases of both healthy and affected individuals will reduce the number of copy number variants of unknown clinical significance and will make genome-wide arrays more useful in clinical practice (ACOG, 2009).

4. Array CGH in prenatal diagnosis

Using the technique of array CGH in Maternal-Fetal Medicine has recently been tested. In 2004, Schaeffer et al. (Schaeffer et al., 2004) observed an increase in detection rate of chromosomal abnormalities not identified by the conventional technique in 9.8% of 41 cases of spontaneous abortions. In frozen lung tissue of 49 fetuses with multiple malformations and normal karyotyping that developed spontaneous abortion or elective termination of pregnancy, Le Caignec et al. (Le Caignec et al., 2005), identified increased detection rate of 16.3% using the technique of array CGH. Another retrospective study (Rickman et al., 2006) has validated two array CGH platforms in 30 samples of amniotic fluid and chorionic villi.

After initial studies that introduced retrospective validation of the technique of array CGH for prenatal diagnosis, the first prospective studies have been emerged. In recent publications, Van den Veyver et al. (Van den Veyver et al., 2009) found 4.8% abnormal findings in 84 fetuses with malformations of the ultrasound, and Kleeman et al. (Kleeman et al., 2009) found the percentage of 8% considering only those fetuses with normal conventional karyotype, showing that the technique of array CGH is a promising tool in prenatal diagnosis. Several reports have now shown the potential utility of array CGH in prenatal diagnosis (Le Caignec et al., 2005, Sahoo et al., 2006, Shaffer et al., 2008, Kleeman et al., 2009, Vialard et al., 2009, Van den Veyver et al., 2009, Machado et al., 2010a).

In the first prospective study in prenatal diagnosis, Sahoo et al. (Sahoo et al., 2006) reported a detection rate of chromosomal abnormalities of 43% of the 98 studied fetuses. Subsequent studies could not found the same rate. In an unselected group of 50 malformed fetuses with normal karyotype, Kleeman et al. (Kleeman et al., 2009) found 4 fetuses (8%) with abnormal array results. Studying a comparable number of cases, by including only fetuses with at least three anomalies and a target array of 287 clones, Le Caignec et al. (Le Caignec et al., 2005) found array abnormalities in 16.3% of the 49 investigated fetuses. Using the same target array CGH to study retrospectively 37 malformed fetuses with at least two anomalies and normal karyotype, Vialard et al. (Vialard et al., 2009) found abnormal results in 4 fetuses corresponding to 10.8%. In Machado et al.(Machado et al., 2010a), although the number of fetuses with copy number changes was much higher than other reports, the number of fetuses with described CNVs among the detected genomic imbalances failed to show such a difference (Table 1).

Study	N of included fetuses	Fetuses with chromosomal imbalances	Fetuses with CNV
Le Caignec et al. (2005)	49	8 (16%)	NL
Sahoo et al. (2006)	98	42 (43%)	30 (71%)
Shaffer et al. (2008)	151*	15 (10%)	12 (80%)
Kleeman et al. (2009)	50	4 (8%)	3 (75%)
Vialard et al. (2009)	37**	4 (10%)	NL
Van den Veyver et al. (2009)	300	58 (19%)	40 (69%)
Machado et al. (2010)	48	45 (94%)	39 (87%)

* Considering only prenatal specimens
** Considering only fetuses with normal karyotype
NL = not listed

Table 1. Number of fetuses with copy number abnormalities and copy number variation (CNV) in recent prenatal array CGH studies

Regardless of array CGH techniques have a theoretical principle and common molecular approaches, published studies have different methodologies in the selection of fetuses, the study design, array CGH platforms, bioinformatics analysis, and interpretation of results, making the comparison between the different works not entirely free from errors. It can explain the difference in detection rates observed in the different studies.

A difference with significant impact on the comparison of results is presented in the selection of fetuses, ranging from fetuses that died with more than three different indicated malformations in prenatal diagnosis, such as maternal anxiety. One study included 367 different indications for molecular analysis, including 20 cases where the indication was not specified. In Machado et. al.(Machado et al., 2010a), the fetuses' selection included only congenital malformations with a strong genetic background, and fetuses with multiple congenital malformations. In both situations is expected a high prevalence of genetic abnormalities and it can explain the high detection rate of fetuses with copy number imbalances in that study.

There is still no consensus among research groups even for interpretation of the results. The definition of "results of uncertain significance" is not clear in the methodology adopted by some of the studies. It can happen due to very little is known about the natural history and range of clinical variability associated with recently described submiscroscopic deletions and duplications detected by array CGH. It is worth noting that the presence of a deletion or duplication alone does not necessarily mean that the copy number alteration causes the observed phenotype and we can not also assure to consider a copy number imbalance as to be pathogenic on the basis only of the association with fetal malformation indentified by ultrasound examination. However, the fetus presenting a significant structural anomaly has a high *a priori* risk of having a pathogenic genetic abnormality and, as is true for any test, the found copy number imbalance is more likely to be a true positive pathogenic one.

In evaluating the scope and content of the regions affected by genomic alterations, Machado et al.(Machado et al., 2010a) identified in 13 of the 48 fetuses (27%) studied by CGH genomic amplifications or deletions that would lead to a modification of the initial cytogenetic results with clinical impact. For these cases, the implications of molecular diagnosis involve experts from the reference to the therapeutic intervention for specific syndromes, to tracking of other malformations. The implications also extend towards reducing the number of diagnostic procedures that the patient could be subjected. For the family, the diagnosis can decrease anxiety and allow for an adequate risk counseling and planning for future offspring. Cytogenetic changes considered clinically significant and ranged from 80Kb to 30Mb in size and were mostly genomic losses.

The clinical applicability of array CGH technique in prenatal diagnosis seems well established for refining diagnosis in cases with suspicious or inconclusive diagnosis of chromosomal structural changes. In recent literature, we find an increasing number of descriptions of cases or series that reinforce this clinical applicability prenatal array CGH (Machado et al., 2010b, Kitsiou-Tzeli et al., 2008). Other situations, for instance, include the study of supposedly balanced translocations by conventional karyotyping, but revealed his unbalanced pattern to the array CGH (Simovich et al., 2007) and the exact sizing and location of chromosome structural abnormalities.

Other clinical application can be checked using the fetuses' selection according to specific defect, as demonstrated for the molecular characterization of fetuses with holoprosencephaly (Machado et al., 2011a) and congenital diaphragmatic hernia.(Machado et al., 2011b) The array CGH could contribute to the knowledge of the submicroscopic genomic instability characterization of specific congenital abnormalities. The indicated significant chromosomal regions are supported when considered that a copy number imbalance in such region was recurrent in fetuses with the same phenotype and when the same genotype-phenotype correlation has been already described. This way, it could identify some clones with uncertain but putative significance that provided a list of chromosomal regions of clinical interest for further molecular evaluation. Additional and confirmatory researches are needed to further establish the role of genes from this chromosome region in the pathogenesis of each specific congenital defect.

The identification of CNVs can greatly complicate the interpretation of the results of array CGH techniques. Among fetuses with changes in the number of copies, 70 to 87% had at least one clone that contained altered, in whole or in part, a chromosomal region with copy number variation (CNV) described in the available databases (see Table 1). This issue is particularly critical for prenatal diagnosis, where the "normal" or not the result defines the perinatal approach. An adequate discrimination between harmless and the real variations aberrations is essential for proper counseling.

Considering all the above and based on the American College of Obstetrics and Gynecology Statement published in 2009 (ACOG, 2009), the array CGH can not replace classic cytogenetics in prenatal diagnosis. According the ACOG opinion, the usefulness of array CGH as a first-line tool in detecting chromosomal abnormalities in all amniocentesis or chorionic villus samples is still unknown. The additional detection rate of chromosomal abnormalities using array CGH, as compared with conventional karyotype for routine fetal chromosomal analysis, awaits a larger population-based study (Table 2).

ACOG RECOMMENDATIONS: (ACOG, 2009)

» Conventional karyotyping remains the principal cytogenetic tool in prenatal diagnosis.

» Targeted array CGH, in concert with genetic counseling, can be offered as an adjunct tool in prenatal cases with abnormal anatomic findings and a normal conventional karyotype, as well as in cases of fetal demise with congenital anomalies and the inability to obtain a conventional karyotype.

» Couples choosing targeted array CGH should receive both pretest and posttest genetic counseling. Followup
genetic counseling is required for interpretation of array CGH results. Couples should understand that array CGH will not detect all genetic pathologies and that array CGH results may be difficult to interpret.

» Targeted array CGH may be useful as a screening tool; however, further studies are necessary to fully determine its utility and its limitations.

Table 2. ACOG recommendations for array CGH in prenatal diagnosis.

The International Standard Cytogenomic Array (ISCA) Consortium, an international group of experts in the array CGH field, held two international workshops and conducted a literature review of 33 studies, including 21,698 patients tested by chromosomal microarray. They provided an evidence-based summary of clinical cytogenetic testing comparing CMA to G-banded karyotyping with respect to technical advantages and limitations, diagnostic yield for various types of chromosomal aberrations, and issues that affect test interpretation. They concluded that the available evidence strongly supports the use of CMA in place of G-banded karyotyping as the first-tier cytogenetic diagnostic test for patients with developmental delay/intellectual disability, autism spectrum disorders or multiple congenital anomalies MCA. However, the ISCA Consortium recognizes that current evidence is not sufficient to allow recommendations regarding prenatal multiple congenital anomalies, and traditional cytogenetic methods, such as G-banded karyotyping and fluorescence in situ hybridization (FISH), are still the standard for prenatal diagnosis (Miller et al., 2010).

5. Challenges

It is recognized that array CGH can detect submicroscopic changes that can be missed on routine chromosomal analysis, especially in prenatal samples where the band resolution may be compromised. However, the exact role of the array CGH in the flowchart of prenatal diagnosis has not been established. Multicenter studies making a direct comparison of the performance of array CGH technique to conventional cytogenetic analysis in a prenatal setting are needed. So far, there's no evidence that it may replace conventional G-banded karyotype analysis, but it can complement and expand current methods for a precise prenatal diagnosis and syndromes' characterization.

The availability of the recent array CGH platforms for fetal chromosomal investigation, including the oligonucleotide arrays (Bi et al., 2008), has brought other challenge controversies over the ideal genetic testing for prenatal diagnosis (Pergament, 2007, Ogilvie et al., 2009, Friedman, 2009). Until now, the oligonucleotide array CGH has not been proven beneficial in relation to protocols with BAC clones for prenatal diagnosis (Bi et al., 2008). Many of the known genomic disorders that can be detected on target arrays do not show readily detectable fetal abnormalities on prenatal ultrasound examinations. Ordering array CGH only in the presence of ultrasound abnormalities may limit the diagnostic potential of this assay (Van den Veyver et al., 2009). Moreover, additional research is needed to further reach a consensus on the optimum platform of an array for clinical use in prenatal diagnosis. It's desirable that, in the future, customized chips with markers across loci discovered should be designed for prenatal diagnosis.

Also, further studies are needed to validate the clinical application of CGH arrays for different clinical situations of prenatal diagnosis, such as advanced maternal age, biochemical screening changed, first-trimester sonographic markers and change in situations of family anxiety in the presence of normal biochemical or ultrasound screening (Pergament, 2007).

Another challenge to overcome is to design a uniform and effective strategy for interpretation of results involving the copy-number variants. The time and effort required for distinguishing the pathogenic and benign findings increases as the resolution of the array CGH increases, but uninterpretable results occur with all array CGH platforms (Friedman, 2009).

Despite the challenges still to overcome, it is clear that the array-based CGH has been asserting itself as a valuable tool in the identification and molecular characterization of chromosomal abnormalities in fetuses with birth defects, opening a new chapter in the historical interface between the Cytogenetics and the Fetal Medicine.

6. References

ACOG 2009. ACOG Committee Opinion No. 446: array comparative genomic hybridization in prenatal diagnosis. *Obstet Gynecol*, 114: 1161-3.

Albertson, D & Pinkel, D 2003. Genomic microarrays in human genetic disease and cancer. *Hum Mol Genet*, 12 Spec No 2: R145-52.

Aradhya, S, Manning, M, Splendore, A & Cherry, A 2007. Whole-genome array-CGH identifies novel contiguous gene deletions and duplications associated with developmental delay, mental retardation, and dysmorphic features. *Am J Med Genet A*, 143A: 1431-41.

Baldwin, E, Lee, J, Blake, D, Bunke, B, Alexander, C, Kogan, A, Ledbetter, D & Martin, C 2008. Enhanced detection of clinically relevant genomic imbalances using a targeted plus whole genome oligonucleotide microarray. *Genet Med*, 10: 415-29.

Ballif, B, Rorem, E, Sundin, K, Lincicum, M, Gaskin, S, Coppinger, J, Kashork, C, Shaffer, L & Bejjani, B 2006. Detection of low-level mosaicism by array CGH in routine diagnostic specimens. *Am J Med Genet A*, 140: 2757-67.

Bejjani, B, Saleki, R, Ballif, B, Rorem, E, Sundin, K, Theisen, A, Kashork, C & Shaffer, L 2005. Use of targeted array-based CGH for the clinical diagnosis of chromosomal imbalance: is less more? *Am J Med Genet A*, 134: 259-67.

Bi, W, Breman, A, Venable, S, Eng, P, Sahoo, T, Lu, X, Patel, A, Beaudet, A, Cheung, S & White, L 2008. Rapid prenatal diagnosis using uncultured amniocytes and oligonucleotide array CGH. *Prenat Diagn*, 28: 943-9.

Boue, A, Muller, F, Briard, M & Boue, J 1988. Interest of biology in the management of pregnancies where a fetal malformation has been detected by ultrasonography. *Fetal Ther*, 3: 14-23.

Brunetti-Pierri, N, Berg, J, Scaglia, F, Belmont, J, Bacino, C, Sahoo, T, Lalani, S, Graham, B, Lee, B, Shinawi, M, Shen, J, Kang, S, Pursley, A, Lotze, T, Kennedy, G, Lansky-Shafer, S, Weaver, C, Roeder, E, Grebe, T, Arnold, G, Hutchison, T, Reimschisel, T, Amato, S, Geragthy, M, Innis, J, Obersztyn, E, Nowakowska, B, Rosengren, S, Bader, P, Grange, D, Naqvi, S, Garnica, A, Bernes, S, Fong, C, Summers, A, Walters, W, Lupski, J, Stankiewicz, P, Cheung, S & Patel, A 2008. Recurrent reciprocal 1q21.1 deletions and duplications associated with microcephaly or macrocephaly and developmental and behavioral abnormalities. *Nat Genet*, 40: 1466-71.

Bryndorf, T, Kirchhoff, M, Rose, H, Maahr, J, Gerdes, T, Karhu, R, Kallioniemi, A, Christensen, B, Lundsteen, C & Philip, J 1995. Comparative genomic hybridization in clinical cytogenetics. *Am J Hum Genet*, 57: 1211-20.

Cabral, A, Machado, I, Leite, H, Pereira, A & Vitral, Z 2001. Cariótipo Fetal em Líquido Pleural Obtido por Toracocentese. *RBGO*, 23: 243-246.

Carpenter, N 2001. Molecular cytogenetics. *Semin Pediatr Neurol*, 8: 135-46.

Caspersson, T, Farber, S, Foley, G, Kudynowski, J, Modest, E, Simonsson, E, Wagh, U & Zech, L 1968. Chemical differentiation along metaphase chromosomes. *Exp Cell Res*, 49: 219-22.

Cheung, S, Shaw, C, Scott, D, Patel, A, Sahoo, T, Bacino, C, Pursley, A, Li, J, Erickson, R, Gropman, A, Miller, D, Seashore, M, Summers, A, Stankiewicz, P, Chinault, A, Lupski, J, Beaudet, A & Sutton, V 2007. Microarray-based CGH detects chromosomal mosaicism not revealed by conventional cytogenetics. *Am J Med Genet A*, 143A: 1679-86.

Cheung, S, Shaw, C, Yu, W, Li, J, Ou, Z, Patel, A, Yatsenko, S, Cooper, M, Furman, P, Stankiewicz, P, Lupski, J, Chinault, A & Beaudet, A 2005 Jul-Aug. Development and validation of a CGH microarray for clinical cytogenetic diagnosis. *Genet Med*, 7: 422-32.

Coe, B, Lockwood, W, Chari, R & Lam, W 2009. Comparative genomic hybridization on BAC arrays. *Methods Mol Biol*, 556: 7-19.

Cornel, M 2000. Wealth and health in relation to birth defects mortality. *J Epidemiol Community Health*, 54: 644.

De Galan-Roosen, A, Kuijpers, J, Meershoek, A & Van Velzen, D 1998. Contribution of congenital malformations to perinatal mortality. A 10 years prospective regional study in The Netherlands. *Eur J Obstet Gynecol Reprod Biol*, 80: 55-61.

Fan, Y, Siu, V, Jung, J & Xu, J 2000. Sensitivity of multiple color spectral karyotyping in detecting small interchromosomal rearrangements. *Genet Test*, 4: 9-14.

Fodor, SP, Read, JL, Pirrung, MC, Stryer, L, Lu, AT & Solas, D 1991. Light-directed, spatially addressable parallel chemical synthesis. *Science,* 251: 767-73.

Friedman, J 2009. High-resolution array genomic hybridization in prenatal diagnosis. *Prenat Diagn,* 29: 20-8.

Gonen, R, Dar, H & Degani, S 1995. The karyotype of fetuses with anomalies detected by second trimester ultrasonography. *Eur J Obstet Gynecol Reprod Biol,* 58: 153-5.

Gribble, S, Prigmore, E, Burford, D, Porter, K, NG, B, Douglas, E, Fiegler, H, Carr, P, Kalaitzopoulos, D, Clegg, S, Sandstrom, R, Temple, I, Youings, S, Thomas, N, Dennis, N, Jacobs, P, Crolla, J & Carter, N 2005. The complex nature of constitutional de novo apparently balanced translocations in patients presenting with abnormal phenotypes. *J Med Genet,* 42: 8-16.

Guo, Z, Guilfoyle, RA, Thiel, AJ, Wang, R & Smith, LM 1994. Direct fluorescence analysis of genetic polymorphisms by hybridization with oligonucleotide arrays on glass supports. *Nucleic Acids Res,* 22: 5456-65.

Haddad, B, Schröck, E, Meck, J, Cowan, J, Young, H, Ferguson-Smith, M, Du Manoir, S & Ried, T 1998. Identification of de novo chromosomal markers and derivatives by spectral karyotyping. *Hum Genet,* 103: 619-25.

Heinrich, J, Machado, I, Vivas, L, Bianchi, M, Cursino Andrade, K, Sbragia, L & Barini, R 2007. Prenatal genomic profiling of abdominal wall defects through comparative genomic hybridization: perspectives for a new diagnostic tool. *Fetal Diagn Ther,* 22: 361-4.

Ishkanian, A, Malloff, C, Watson, S, Deleeuw, R, Chi, B, Coe, B, Snijders, A, Albertson, D, Pinkel, D, Marra, M, Ling, V, Macaulay, C & Lam, W 2004. A tiling resolution DNA microarray with complete coverage of the human genome. *Nat Genet,* 36: 299-303.

Jackson, L 2002. Cytogenetics and molecular cytogenetics. *Clin Obstet Gynecol,* 45: 622-39; discussion 730-2.

Jacobs, P, Browne, C, Gregson, N, Joyce, C & White, H 1992. Estimates of the frequency of chromosome abnormalities detectable in unselected newborns using moderate levels of banding. *J Med Genet,* 29: 103-8.

Kallioniemi, A, Kallioniemi, O, Sudar, D, Rutovitz, D, Gray, J, Waldman, F & Pinkel, D 1992. Comparative genomic hybridization for molecular cytogenetic analysis of solid tumors. *Science,* 258: 818-21.

Kalter, H & Warkany, J 1983. Medical progress. Congenital malformations: etiologic factors and their role in prevention (first of two parts). *N Engl J Med,* 308: 424-31.

Khrapko, KR, Lysov Yup, Khorlyn, AA, Shick, VV, Florentiev, VL & Mirzabekov, AD 1989. An oligonucleotide hybridization approach to DNA sequencing. *FEBS Lett,* 256: 118-22.

Kirchhoff, M, Gerdes, T, Maahr, J, Rose, H, Bentz, M, Döhner, H & Lundsteen, C 1999. Deletions below 10 megabasepairs are detected in comparative genomic hybridization by standard reference intervals. *Genes Chromosomes Cancer,* 25: 410-3.

Kirchhoff, M, Rose, H & Lundsteen, C 2001. High resolution comparative genomic hybridisation in clinical cytogenetics. *J Med Genet,* 38: 740-4.

Kitsiou-Tzeli, S, Sismani, C, Karkaletsi, M, Florentin, L, Anastassiou, A, Koumbaris, G, Enangelidou, P, Agapitos, E, Patsalis, P & Velissariou, V 2008. Prenatal diagnosis of a de novo partial trisomy 10p12.1-12.2 pter originating from an unbalanced translocation onto 15qter and confirmed with array CGH. *Prenat Diagn,* 28: 770-2.

Kleeman, L, Bianchi, D, Shaffer, L, Rorem, E, Cowan, J, Craigo, S, Tighiouart, H & Wilkins-Haug, L 2009. Use of array comparative genomic hybridization for prenatal diagnosis of fetuses with sonographic anomalies and normal metaphase karyotype. *Prenat Diagn,* 29: 1213-7.

Lamture, JB, Beattie, KL, Burke, BE, Eggers, MD, Ehrlich, DJ, Fowler, R, Hollis, MA, Kosicki, BB, Reich, RK & Smith, SR 1994. Direct detection of nucleic acid hybridization on the surface of a charge coupled device. *Nucleic Acids Res,* 22: 2121-5.

Lapierre, J, Cacheux, V, Collot, N, Da Silva, F, Hervy, N, Rivet, D, Romana, S, Wiss, J, Benzaken, B, Aurias, A & Tachdjian, G 1998. Comparison of comparative genomic hybridization with conventional karyotype and classical fluorescence in situ hybridization for prenatal and postnatal diagnosis of unbalanced chromosome abnormalities. *Ann Genet,* 41: 133-40.

Le Caignec, C, Boceno, M, Saugier-Veber, P, Jacquemont, S, Joubert, M, David, A, Frebourg, T & Rival, J 2005. Detection of genomic imbalances by array based comparative genomic hybridisation in fetuses with multiple malformations. *J Med Genet,* 42: 121-8.

Lu, X, Shaw, C, Patel, A, LI, J, Cooper, M, Wells, W, Sullivan, C, Sahoo, T, Yatsenko, S, Bacino, C, Stankiewicz, P, OU, Z, Chinault, A, Beaudet, A, Lupski, J, Cheung, S & Ward, P 2007. Clinical implementation of chromosomal microarray analysis: summary of 2513 postnatal cases. *PLoS One,* 2: e327.

Lucito, R, Healy, J, Alexander, J, Reiner, A, Esposito, D, Chi, M, Rodgers, L, Brady, A, Sebat, J, Troge, J, West, J, Rostan, S, Nguyen, K, Powers, S, YE, K, Olshen, A, Venkatraman, E, Norton, L & Wigler, M 2003. Representational oligonucleotide microarray analysis: a high-resolution method to detect genome copy number variation. *Genome Res,* 13: 2291-305.

Machado, I, Heinrich, J & Barini, R 2010a. Whole genome BAC-array CGH results in fetuses with congenital malformation and normal karyotype. In *Advances in Perinatal Medicine.,* Puertas, A, Montoya, F, Romero, J, Hurtado, J & Manzanares, S (ed.)^(eds.). Monduzzi Editore: Granada, Spain.

Machado, IN, Heinrich, JK & Barini, R 2011a. Genomic imbalances detected through array CGH in fetuses with holoprosencephaly. *Arq Neuropsiquiatr,* 69: 3-8.

Machado, IN, Heinrich, JK, Barini, R & Peralta, CF 2011b. Copy number imbalances detected with a BAC-based array comparative genomic hybridization platform in congenital diaphragmatic hernia fetuses. *Genet Mol Res,* 10: 261-7.

Machado, IN, Heinrich, JK, Campanhol, C, Rodrigues-Peres, RM, Oliveira, FM & Barini, R 2010b. Prenatal diagnosis of a partial trisomy 13q (q14-->qter): phenotype, cytogenetics and molecular characterization by spectral karyotyping and array comparative genomic hybridization. *Genet Mol Res,* 9: 441-8.

Mantripragada, K, Tapia-Páez, I, Blennow, E, Nilsson, P, Wedell, A & Dumanski, J 2004. DNA copy-number analysis of the 22q11 deletion-syndrome region using array-CGH with genomic and PCR-based targets. *Int J Mol Med,* 13: 273-9.

Maskos, U & Southern, EM 1992. Oligonucleotide hybridizations on glass supports: a novel linker for oligonucleotide synthesis and hybridization properties of oligonucleotides synthesised in situ. *Nucleic Acids Res,* 20: 1679-84.

Menten, B, Maas, N, Thienpont, B, Buysse, K, Vandesompele, J, Melotte, C, De Ravel, T, Van Vooren, S, Balikova, I, Backx, L, Janssens, S, De Paepe, A, De Moor, B, Moreau, Y, Marynen, P, Fryns, J, Mortier, G, Devriendt, K, Speleman, F & Vermeesch, J 2006. Emerging patterns of cryptic chromosomal imbalance in patients with idiopathic mental retardation and multiple congenital anomalies: a new series of 140 patients and review of published reports. *J Med Genet,* 43: 625-33.

Miller, DT, Adam, MP, Aradhya, S, Biesecker, LG, Brothman, AR, Carter, NP, Church, DM, Crolla, JA, Eichler, EE, Epstein, CJ, Faucett, WA, Feuk, L, Friedman, JM, Hamosh, A, Jackson, L, Kaminsky, EB, Kok, K, Krantz, ID, Kuhn, RM, Lee, C, Ostell, JM,

Rosenberg, C, Scherer, SW, Spinner, NB, Stavropoulos, DJ, Tepperberg, JH, Thorland, EC, Vermeesch, JR, Waggoner, DJ, Watson, MS, Martin, CL & Ledbetter, DH 2010. Consensus statement: chromosomal microarray is a first-tier clinical diagnostic test for individuals with developmental disabilities or congenital anomalies. *Am J Hum Genet,* 86: 749-64.

Nelson, K & Holmes, L 1989. Malformations due to presumed spontaneous mutations in newborn infants. *N Engl J Med,* 320: 19-23.

Ness, G, Lybaek, H & Houge, G 2002. Usefulness of high-resolution comparative genomic hybridization (CGH) for detecting and characterizing constitutional chromosome abnormalities. *Am J Med Genet,* 113: 125-36.

Nicolaides, K, Snijders, R, Gosden, C, Berry, C & Campbell, S 1992. Ultrasonographically detectable markers of fetal chromosomal abnormalities. *Lancet,* 340: 704-7.

Nussbaum, R, Mcinnes, R & Willard, H 2008. *Thompson & Thompson Genética Médica.* Elsevier.: Philadelphia.

Ogilvie, C, Yaron, Y & Beaudet, A 2009. Current controversies in prenatal diagnosis 3: For prenatal diagnosis, should we offer less or more than metaphase karyotyping? *Prenat Diagn,* 29: 11-4.

Pergament, E 2007. Controversies and challenges of array comparative genomic hybridization in prenatal genetic diagnosis. *Genet Med,* 9: 596-9.

Pickering, D, Eudy, J, Olney, A, Dave, B, Golden, D, Stevens, J & Sanger, W 2008. Array-based comparative genomic hybridization analysis of 1176 consecutive clinical genetics investigations. *Genet Med,* 10: 262-6.

Pinkel, D, Segraves, R, Sudar, D, Clark, S, Poole, I, Kowbel, D, Collins, C, Kuo, W, Chen, C, Zhai, Y, Dairkee, S, Ljung, B, Gray, J & Albertson, D 1998. High resolution analysis of DNA copy number variation using comparative genomic hybridization to microarrays. *Nat Genet,* 20: 207-11.

Pollack, J, Perou, C, Alizadeh, A, Eisen, M, Pergamenschikov, A, Williams, C, Jeffrey, S, Botstein, D & Brown, P 1999. Genome-wide analysis of DNA copy-number changes using cDNA microarrays. *Nat Genet,* 23: 41-6.

Re, S 1993. Causes of congenital anomalies: an overview and historical perspective. In *Human malformations and related anomalies,* Stevenson RE, HJ, Goodman RM (ed.)^(eds.). Oxford University Press: New York.

Rickman, L, Fiegler, H, Shaw-Smith, C, Nash, R, Cirigliano, V, Voglino, G, Ng, B, Scott, C, Whittaker, J, Adinolfi, M, Carter, N & Bobrow, M 2006. Prenatal detection of unbalanced chromosomal rearrangements by array CGH. *J Med Genet,* 43: 353-61.

Rosano, A, Botto, L, Botting, B & Mastroiacovo, P 2000. Infant mortality and congenital anomalies from 1950 to 1994: an international perspective. *J Epidemiol Community Health,* 54: 660-6.

Rosenberg, C, Knijnenburg, J, Bakker, E, Vianna-Morante, A, Sloos, W, Otto, P, Kriek, M, Hansson, K, Krepischi-Santos, A, Fiegler, H, Carter, N, Bijlsma, E, Van Haeringen, A, Szuhai, K & Tanke, H 2006. Array-CGH detection of micro rearrangements in mentally retarded individuals: clinical significance of imbalances present both in affected children and normal parents. *J Med Genet,* 43: 180-6.

Ryall, R, Callen, D, Cocciolone, R, Duvnjak, A, Esca, R, Frantzis, N, Gjerde, E, Haan, E, Hocking, T, Sutherland, G, Thomas, D & Webb, F 2001. Karyotypes found in the population declared at increased risk of Down syndrome following maternal serum screening. *Prenat Diagn,* 21: 553-7.

Sahoo, T, Cheung, S, Ward, P, Darilek, S, Patel, A, Del Gaudio, D, Kang, S, Lalani, S, Li, J, Mcadoo, S, Burke, A, Shaw, C, Stankiewicz, P, Chinault, A, Van Den Veyver, I, Roa, B, Beaudet, A & Eng, C 2006. Prenatal diagnosis of chromosomal abnormalities using array-based comparative genomic hybridization. *Genet Med,* 8: 719-27.

Schaeffer, A, Chung, J, Heretis, K, Wong, A, Ledbetter, D & Lese Martin, C 2004. Comparative genomic hybridization-array analysis enhances the detection of aneuploidies and submicroscopic imbalances in spontaneous miscarriages. *Am J Hum Genet,* 74: 1168-74.

Schena, M, Shalon, D, Davis, Rw & Brown, Po 1995. Quantitative monitoring of gene expression patterns with a complementary DNA microarray. *Science,* 270: 467-70.

Schoumans, J, Ruivenkamp, C, Holmberg, E, Kyllerman, M, Anderlid, B & Nordenskjöld, M 2005. Detection of chromosomal imbalances in children with idiopathic mental retardation by array based comparative genomic hybridisation (array-CGH). *J Med Genet,* 42: 699-705.

Schröck, E, Veldman, T, Padilla-Nash, H, Ning, Y, Spurbeck, J, Jalal, S, Shaffer, L, Papenhausen, P, Kozma, C, Phelan, M, Kjeldsen, E, Schonberg, S, O'brien, P, Biesecker, L, Du Manoir, S & Ried, T 1997. Spectral karyotyping refines cytogenetic diagnostics of constitutional chromosomal abnormalities. *Hum Genet,* 101: 255-62.

Shaffer, L, Coppinger, J, Alliman, S, Torchia, B, Theisen, A, Ballif, B & Bejjani, B 2008. Comparison of microarray-based detection rates for cytogenetic abnormalities in prenatal and neonatal specimens. *Prenat Diagn,* 28: 789-95.

Shaffer, L, Kashork, C, Saleki, R, Rorem, E, Sundin, K, Ballif, B & Bejjani, B 2006. Targeted genomic microarray analysis for identification of chromosome abnormalities in 1500 consecutive clinical cases. *J Pediatr,* 149: 98-102.

Shaikh, T 2007. Oligonucleotide arrays for high-resolution analysis of copy number alteration in mental retardation/multiple congenital anomalies. *Genet Med,* 9: 617-25.

Sharkey, F, Maher, E & Fitzpatrick, D 2005. Chromosome analysis: what and when to request. *Arch Dis Child,* 90: 1264-9.

Sharp, A, Locke, D, Mcgrath, S, Cheng, Z, Bailey, J, Vallente, R, Pertz, L, Clark, R, Schwartz, S, Segraves, R, Oseroff, V, Albertson, D, Pinkel, D & Eichler, E 2005. Segmental duplications and copy-number variation in the human genome. *Am J Hum Genet,* 77: 78-88.

Shaw, C & Lupski, J 2004. Implications of human genome architecture for rearrangement-based disorders: the genomic basis of disease. *Hum Mol Genet,* 13 Spec No 1: R57-64.

Shaw-Smith, C, Redon, R, Rickman, L, Rio, M, Willatt, L, Fiegler, H, Firth, H, Sanlaville, D, Winter, R, Colleaux, L, Bobrow, M & Carter, N 2004. Microarray based comparative genomic hybridisation (array-CGH) detects submicroscopic chromosomal deletions and duplications in patients with learning disability/mental retardation and dysmorphic features. *J Med Genet,* 41: 241-8.

Shinawi, M, Shao, L, Jeng, L, Shaw, C, Patel, A, Bacino, C, Sutton, V, Belmont, J & Cheung, S 2008. Low-level mosaicism of trisomy 14: phenotypic and molecular characterization. *Am J Med Genet A,* 146A: 1395-405.

Simovich, M, Yatsenko, S, Kang, S, Cheung, S, Dudek, M, Pursley, A, Ward, P, Patel, A & Lupski, J 2007. Prenatal diagnosis of a 9q34.3 microdeletion by array-CGH in a fetus with an apparently balanced translocation. *Prenat Diagn,* 27: 1112-7.

Smith, L, Nagar, S, Kim, G & Morgan, W 2003. Radiation-induced genomic instability: radiation quality and dose response. *Health Phys,* 85: 23-9.

Solinas-Toldo, S, Lampel, S, Stilgenbauer, S, Nickolenko, J, Benner, A, Döhner, H, Cremer, T & Lichter, P 1997. Matrix-based comparative genomic hybridization: biochips to screen for genomic imbalances. *Genes Chromosomes Cancer,* 20: 399-407.

Speicher, M, Gwyn Ballard, S & Ward, D 1996. Karyotyping human chromosomes by combinatorial multi-fluor FISH. *Nat Genet,* 12: 368-75.

Stankiewicz, P & Beaudet, A 2007. Use of array CGH in the evaluation of dysmorphology, malformations, developmental delay, and idiopathic mental retardation. *Curr Opin Genet Dev,* 17: 182-92.

Telenius, H, Carter, N, Bebb, C, Nordenskjöld, M, Ponder, B & Tunnacliffe, A 1992. Degenerate oligonucleotide-primed PCR: general amplification of target DNA by a single degenerate primer. *Genomics,* 13: 718-25.

Thein, A, Charles, A, Davies, T, Newbury-Ecob, R & Soothill, P 2001. The role of comparative genomic hybridisation in prenatal diagnosis. *BJOG,* 108: 642-8.

Van Den Veyver, I, Patel, A, Shaw, C, Pursley, A, Kang, S, Simovich, M, Ward, P, Darilek, S, Johnson, A, Neill, S, Bi, W, White, L, Eng, C, Lupski, J, Cheung, S & Beaudet, A 2009. Clinical use of array comparative genomic hybridization (aCGH) for prenatal diagnosis in 300 cases. *Prenat Diagn,* 29: 29-39.

Vermeesch, J, Melotte, C, Froyen, G, Van Vooren, S, Dutta, B, Maas, N, Vermeulen, S, Menten, B, Speleman, F, De Moor, B, Van Hummelen, P, Marynen, P, Fryns, J & Devriendt, K 2005. Molecular karyotyping: array CGH quality criteria for constitutional genetic diagnosis. *J Histochem Cytochem,* 53: 413-22.

Vialard, F, Molina Gomes, D, Leroy, B, Quarello, E, Escalona, A, Le Sciellour, C, Serazin, V, Roume, J, Ville, Y, De Mazancourt, P & Selva, J 2009. Array comparative genomic hybridization in prenatal diagnosis: another experience. *Fetal Diagn Ther,* 25: 277-84.

Vissers, L, De Vries, B, Osoegawa, K, Janssen, I, Feuth, T, Choy, C, Straatman, H, Van Der Vliet, W, Huys, E, Van Rijk, A, Smeets, D, Van Ravenswaaij-Arts, C, Knoers, N, Van Der Burgt, I, De Jong, P, Brunner, H, Van Kessel, A, Schoenmakers, E & Veltman, J 2003. Array-based comparative genomic hybridization for the genomewide detection of submicroscopic chromosomal abnormalities. *Am J Hum Genet,* 73: 1261-70.

Vissers, L, Veltman, J, Van Kessel, A & Brunner, H 2005. Identification of disease genes by whole genome CGH arrays. *Hum Mol Genet,* 14 Spec No. 2: R215-23.

Weiss, M, Kuipers, E, Meuwissen, S, Van Diest, P & Meijer, G 2003a. Comparative genomic hybridisation as a supportive tool in diagnostic pathology. *J Clin Pathol,* 56: 522-7.

Weiss, M, Kuipers, E, Postma, C, Snijders, A, Siccama, I, Pinkel, D, Westerga, J, Meuwissen, S, Albertson, D & Meijer, G 2003b. Genomic profiling of gastric cancer predicts lymph node status and survival. *Oncogene,* 22: 1872-9.

Wladimiroff, J, Sachs, E, Reuss, A, Stewart, P, Pijpers, L & Niermeijer, M 1988. Prenatal diagnosis of chromosome abnormalities in the presence of fetal structural defects. *Am J Med Genet,* 29: 289-91.

Wood, E, Dowey, S, Saul, D, Cain, C, Rossiter, J, Blakemore, K & Stetten, G 2008. Prenatal diagnosis of mosaic trisomy 8q studied by ultrasound, cytogenetics, and array-CGH. *Am J Med Genet A,* 146A: 764-9.

Yu, L, Moore, Dn, Magrane, G, Cronin, J, Pinkel, D, Lebo, R & Gray, J 1997. Objective aneuploidy detection for fetal and neonatal screening using comparative genomic hybridization (CGH). *Cytometry,* 28: 191-7.

Zeitune, M, Aitken, D, Crossley, J, Yates, J, Cooke, A & Ferguson-Smith, M 1991. Estimating the risk of a fetal autosomal trisomy at mid-trimester using maternal serum alpha-fetoprotein and age: a retrospective study of 142 pregnancies. *Prenat Diagn,* 11: 847-57.

6

Genetic Studies in Acute Lymphoblastic Leukemia, from Diagnosis to Optimal Patient's Treatment

Małgorzata Krawczyk-Kuliś
Medical University of Silesia
Poland

1. Introduction

In the recent years progress in the basic laboratory science has allowed for implementation of many advanced methods in the clinical practice. Thereupon French- American- British Group (FAB) classification of acute leukemias and the immunological classification, published in 1995 by European Group for the Immunological Characterization of Leukemias (Bene, 1995), were changed into WHO 2008 classification, which applied genetic investigations to differentiate subtypes of acute leukemia (WHO, 2008). In the former classifications, lymphoma and leukemia were diagnosed as distinct disorders. Apart from including lymphoma and leukemia in the WHO 2008 classification as a single disease, acute lymphoblastic leukemia (ALL) can be diagnosed if over 20% (in some investigations over 25%) lymphoblast infiltration is detected in bone marrow biopsy. In this new classifications precursor lymphoid neoplasms are divided into the B (about 80% of cases) or T cell lineage (about 15-25%). ALL diagnosis, based on antigen B or T investigation, uses flow cytometry method. Recent investigations suggest different molecular profiles for ALL-T type and T-lymphoblastic lymphoma especially an expression of CD47 in T -ALL, and over-expression of *MLL1* in T- lymphoblastic lymphoma (Hoelzer & Gokbuget, 2009; Raetz et al., 2006). The prevalence of ALL in children amounts to 30-35% of neoplastic diseases and its incidence is approximately 40 cases in a million per year. In adults ALL account for about 20% of all types of acute leukemia. The incidence in adults is estimated as 0,39 per 100 000 per year in 35-39 age range, and increases to 2,1 patients over 80 years (Anino et al., 2002). The distinction of various ALL subtypes characterized by recurrent genetic abnormalities was made possible thanks to specific genetic studies. The incidence of the diagnosed subtypes occurring in adult and children population varies and indicates differences in the clinical features (Harrison, 2008). Cytogenetic studies have become a routine procedure in clinical practice involved in acute leukemia treatment (Faderl et al., 1998). Carrying out of the above-mentioned methods is an indispensable condition to make a proper diagnosis according to WHO 2008 classification.

Cytogenetic studies implicate specific types of therapy in adults and children as well (Tomizawa et al., 2007).

However, the results of treatment of ALL patients improved not only because of better treatment modality standards and facilities but also thanks to proper and detailed diagnosis (Faderl et al., 2010).

I.	1. B lymphoblastic leukemia/lymphoma, not otherwise specified
B lymphoblastic leukemia/lymphoma	2. B lymphoblastic leukemia/lymphoma with recurrent genetic abnormalities
	- B lymphoblastic leukemia/lymphoma with t(9;22)(q34;q11.2); *BCR-ABL*1
	- B lymphoblastic leukemia/lymphoma with t(v;11q23);*MLL* rearranged
	- B lymphoblastic leukemia/lymphoma with t(12;21)(p13;q22);*TEL-AML*1 (*ETV6-RUNX*1)
	- B lymphoblastic leukemia/lymphoma with hyperdiploidy
	- B lymphoblastic leukemia/lymphoma with hypodiploidy (Hypodiploid ALL)
	- B lymphoblastic leukemia/lymphoma with t(5;14)(q31;q32);*IL3-IGH*
	- B lymphoblastic leukemia/lymphoma with t(1;19)(q23;p13.3);E2A-*PBX*1(*TCF3-PBX*1)
II. T lymphoblastic leukemia/lymphoma	

Table 1. Precursor Lymphoid Neoplasms Classification WHO 2008.

2. Flow cytometry

Flow cytometry is a method, which is most commonly used for clinical diagnosis of 'de novo' acute leukemias, and as such does not require any additional preparation on the part of the patient. It is the method of sorting and measuring types of cells by fluorescent labelling of monoclonal antibodies on the surface or in cytoplasm of investigated cells. Types of an antigen or other markers present on the cell give further information about the immunophenotype of leukemic cells. In the method monoclonal antibodies are used for detecting antigens determined in CD classification. Despite that, it is not the method of cytogenetic examination; the frequent application of immunophenotyping to peripheral blood or bone marrow aspiration cells necessitates the description of the flow cytometry in this chapter too.

Flow cytometry investigation of neoplastic cells is commonly used for diagnosis of central nervous infiltration of ALL manifested as leptomeningeal disease. The National Comprehensive Cancer Network recommends the routine use of the flow cytometry for the diagnosis of the central nervous infiltration involvement in ALL (Brem et al., 2008).

The process of collecting data from samples is performed using a flow cytometer. The data generated by flow cytometers are presented as a 'plot' i.e. histogram. The histogram regions can be sequentially separated, based on fluorescence intensity, by creating a series of subset extractions, called 'gates'. For diagnostic purposes in hematology specific gating protocols exist. The WHO 2008 classification divided leukemias/lymphomas into B or T cell types and these types of leukemia can be distinguished both by immunophenotype and by molecular genetic studies.

The B lymphoblasts as well as the T lymphoblasts can express a panel of characteristic antigens (Tab.2).

Basic panel for diagnosis of acute leukemia	Panel of antigens for B derived ALL	Panel of antigens for T derived ALL	Minimal residual disease monitoring
CD 45 CD10 CD1a CD7 CD22 CD65 CD2 CD13 CD14 CD33 CD34 CD117 CD15 CD56 Additionally: CD16, CD66, CD36, CD64, CD 41, GlyA CD11b,CD11c, HLADR, CD38,	CD34 CD19 CD10 CD20 cyCD22 CD38 cyCD79a CD9 CD45 CD45RA IgG2a IgG3 CD52 CD58 Additionally: TdT, sIg, cyIgM, cyIgG1, IgG1, IgM, lambda, kappa For ALL CD19 positive: CD33, CD13, CD117, CD15BD, CD2, CD7, CD66, CD123	CD7 IgG1 CD2 CD3 CD5 CD4 CD8 cy/sCD3 CD1a CD34 CD38 CD45 IgG3 CD52 CD99 Additionally: cyIgG1, cyCD3 CD45RA, CD45RO, CD25, CD57, CD16 For ALL CD7 positive: CD33, CD13, CD117, CD15BD, CD56, CD65, CD117	Case nr 1, (B-cell ALL): Qu*: CD66c/19 and CD10/19/45RA; and CD45/19 or ES**: CD58/19, CD10/20/19, CD34/38/TdT, CD10/19/TdT, CD34/38/TdT, CD10/19/TdT Case nr 2, (B-cell ALL): Qu: Cd34/66c/19, ES: CD10/19/20, CD34/38/19, CD45/34/19, CD34/9/19, CD58/51/19, CD10/19/TdT Case nr 3, (B-cell ALL): CD19/22/34, CD19/TdT, CD33/HLADr Case nr 4, (T-cell ALL): sCD7/TdT/cCD3, CD7+/5+/3 negative, CD7+/1a+/3 negative, CD4+/8+/3 negative

Table 2. The list of monoclonal antibodies for diagnosis and monitoring the treatment of acute lymphoblastic leukemia

The progress in the flow cytometry technique allowed to obtain information about 17-60 parameters of the investigated cells simultaneously (Wood et al., 2006).

This method, which is useful in diagnosis, can additionally detect a very small number of abnormal cells in bone marrow suspension or peripheral blood cells obtained from patients after the treatment, hence can detect the minimal (i.e. submicroscopic) residual disease (MRD) (Campana, 2009). The panel of antigens for MRD examination is matched individually on the basis of the results of immunophenotyping, which is performed during diagnosis. At least two different aberrant phenotypes with the expression over 50% of leukemic blasts are used on the average. The following are necessary and useful criteria in monitoring MRD: coexpression of antigens from different than lymphoblastic cell lines, e.g. CD13, CD33, asynchronous antigens expression or overexpression within the same line and ectopic phenotypes. MRD can be evaluated using either the 'quadrant' method or 'empty spaces' technique. Usually, MRD is calculated as the percentage of total nuclear bone marrow cells and in 3-8 colour flow cytometry method, where the 0,1% -0,01% sensitivity is obtained. Flow cytometry is described as a widely applicable, rapid and accurate quantification method which provides additional information on normal hematopoietic cells and can differentiate these cells from neoplastic compartment (Figure 1). The potential weakness is phenotypic shifts and as a result there are multiple aberrant phenotypes required.

The results of immunophenotyping can be useful in clinical application for identifying antigens in targeted therapy, and as a minimal residual disease monitoring during treatment (Giebel et al., 2010; Rhein et al., 2010). Targeted therapy involves the use of monoclonal antibodies as follows: anti CD20 (rituximab), anti CD52 (alemtuzumab), anti CD22 (epratuzumab), anti CD 33 (+toxin) (gemtuzumab ozogamycin), and Blinatumomab used recently (Nijmeijer et al., 2010; Raetz et al., 2008; Topp et al., 2009; Topp et al., 2011).

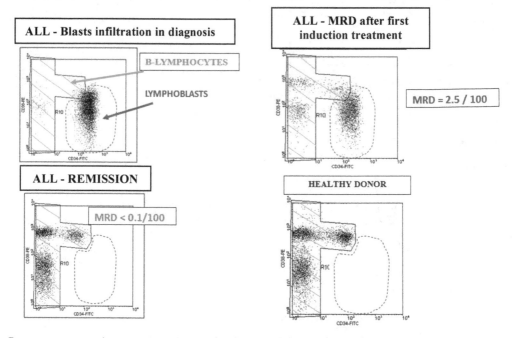

Bone marrow samples were stained using MoAb: CD34-FITC / CD38-PE / CD19-PC5
Diagram 'dot plot' showed only CD19(+) cells
Normal B-Lymphocytes – shaded area
Lymphoblasts – dash area

Fig. 1. Empty Spaces method for MRD monitoring during ALL CD19+ treatment. The figure shows the bone marrow samples examination.

3. Cytogenetic methods used in acute lymphoblastic leukemia

Cytogenetic changes underline leukemogenesis targeted and are closely associated with the type neoplasm developing. Several cases of precursor lymphoid neoplasms have characteristic genetic abnormalities that are important in determining their biologic and clinical features (Harrison, 2001; Mullingan, 2009). Some of mutations occurred both in ALL and in AML, e.g. *FLT3* mutations (point mutations and internal tandem duplications (Chang P., 2010). Chromosomal abnormalities are detected in about 80% of ALL cases but in about 40% numerical abnormalities exist and in 40% there are structural alterations (Witt et al., 2009). Over the years, methods of cytogenetic analysis evolved and became a part of routine laboratory testing, providing valuable diagnostic and prognostic information in children and adult patients (Park et al., 2008). The reference material for cytogenetic investigation the cells obtained during bone marrow aspiration. Peripheral blood cells are used only if bone marrow cells are unavailable or if special methods of cytogenetic investigation are used e.g. fluorescence in situ hybridization or polymerase chain reaction instead of conventional cytogenetics.

3.1 Conventional cytogenetics and fluorescence in situ hybridization

The gold standard for cytogenetical investigation is still conventional cytogenetics (Figure 2) but now often combined with analyses using fluorescence in situ hybridization (FISH), and polymerase chain reaction (PCR) technique.

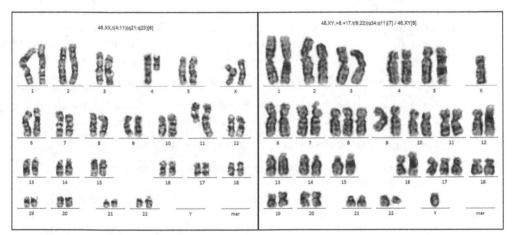

Fig. 2. Results of conventional cytogenetic investigation in ALL patients (GTG method). Images show the t(4;11) and t(9;22).

For cytogenetical investigation the sample of 2-3ml volume from bone marrow aspiration or 5-10ml from peripheral blood, with the addition of heparin, must be delivered to cytogenetic laboratory in 24 hours, and can be transported in a room temperature. This diagnostic procedure must be carried out before the start of any antineoplastic treatment. The main elements of the cytogenetical methods are cell cultures which are mandatory for obtaining metaphases for chromosomal analysis. The duration time of cells cultures oscillates from 0,4-2 hours (called 'immediate') to 24, 48 and 96 hours. Then cytogenetical analysis of cells is performed in metaphases and the rest of cell suspension can be left in minus 20 Celsius degree for additional investigations. According to the guidelines of European Cytogeneticists Association (ECA), prepared as a quality framework for cytogenetic laboratories, diagnostic metaphases must be obtained from at least 90% of cell cultures (Bricarelli et al., 2006). The minimum number of metaphases required to obtain the result, is 20 in normal karyotype. If the karyotype is abnormal the number of metaphases may be lower (sometimes only few metaphases) providing that clonal aberrations are recurring (Haferlach et al., 2007). Cytogenetic findings were reported up to 3 weeks after the sample had been received in laboratory. There are some limitations of conventional cytogenetic analysis such as sometimes morphologically insufficient quality of metaphases or the mitotic index. The paper presents results of investigation of 70 ALL in children revealed, that karyotypes were obtained in 84% (Soszynska et al., 2008).

Fluorescence in situ hybridization method (FISH) is a cytogenetic technique providing detection of characteristic chromosomal DNA sequences, by painting them. FISH is a method of supplementing the classical cytogenetic studies, but can also be an independent method for cytogenetic analysis in ALL. Principle of FISH is the use of DNA fragments precisely defined sequence, that is, molecular probes of complementary hybridizing appropriately prepared DNA test. Special locus-specific probe mixtures are used to count chromosomes. Fluorescence microscopy can be used to find out where the fluorescent sound is bound to the chromosome. To determine the percentage of cells with genetic abnoramalities sought is indicated in the analysis of at least 100 interphase or metaphase cells. There are some modifications of FISH technique that were implemented in hematology laboratories e.g. multiple colour FISH (M-FISH) is widely applied for detection of BCR/ABL translocation and for MLL gene rearrangements. Employed ratios of probe mixtures are supposed to create secondary colours that are useful in differentiating subtypes of ALL.

M-FISH elucidated complex karyotypes (Broadfield et al., 2004; Harrison et al., 1999).

Results of FISH and conventional cytogenetics should be determined in accordance with the International System for Human Cytogenetic Nomenclature (ISCN, 2005).

In ALL diagnostic procedures FISH testing should always be done if:

1. In the conventional cytogenetic study, no metaphases were obtained or the quality is not suitable for release as a result,
2. The outcome of conventional cytogenetics GTG technique suggests the presence of aberrations, but does not confirm it,
3. In conventional cytogenetics was not found chromosome aberrations characteristic for the type of leukemia (e.g. the CD10 positive leukemia with Ph cryptic or masked Ph).

The new technique, which detected chromosomal imbalances, introduced by Kallioniemi (Kallioniemi1 et al., 1994) is comparative genomic hybridization (CGH), which is a method of molecular cytogenetics. CGH gives a global overview of chromosomal deletions and amplifications throughout the whole genome with one step analysis (McGrattan et al., 2008). CGH can detect submicroscopic deletions 5-10Mbp in size and detect extra-chromosomal fragments of chromatin size 2-3Mpz. The major limitation of CGH, however, is the failure to detect balanced chromosome abnormalities such as translocations, inversions, and clonal heterogeneity (Ness et al., 2002). This method can be performed as a complementary test

GTG and FISH. CGH could be established as a routine method of analysis for screening patients with ALL (Kowalczyk et al., 2010).

Test result conventional cytogenetics and FISH help determine the specific subtype of ALL diagnosis, identify risk factors, establishing prognosis, treatment selection, is also to monitor the disease and the effectiveness of treatment.

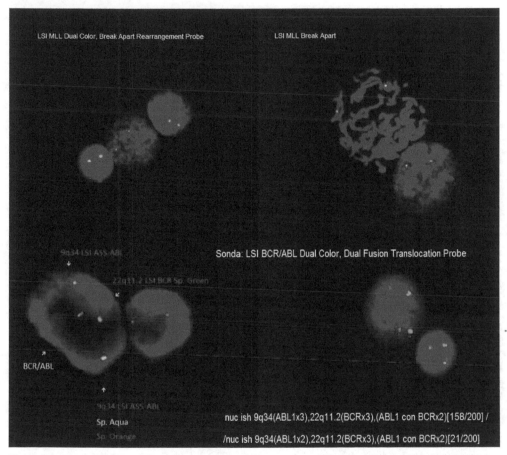

Fig. 3. Results of FISH method detecting translocations in All patients: t(9;22) and t(4;11).

3.2 Polymerase chain reaction

The polymerase chain reaction (PCR), a technique of molecular biology now is a routinely used method for ALL diagnosis as well as for categorizing ALL subtypes according to WHO2008 classification, monitoring the treatment results and measurement of minimal residual disease too.

For PCR investigation the sample of 1-5ml volume from bone marrow aspiration or 5-10ml from peripheral blood, with the addition of EDTA, must be delivered to cytogenetic laboratory in 24 hours, and must be transported in +4ºC.

PCR has a very high sensitivity and detects known fusion events (Wesołowska et al., 2011). Suitably selected set of primers can detect aberrations at the RNA level (the method of RT-PCR). There are some variants of PCR technique which are very useful in hematological investigation e.g. single-nucleotide polymorphisms (SNPs), Asymmetric PCR, Multiplex-PCR, Nested PCR, Quantitative PCR (Q-PCR), Quantitative real-time PCR, Reverse Transcription PCR (RT-PCR). The RT-RQ-PCR method can detect translocation-specific malignant lymphoblasts at a sensitivity determined as 10^{-4} - 10^{-6}. The method is used for monitoring MRD and its advantage is the stability of gene fusion in the course of the disease, and high sensitivity determinations. The disadvantage of RT-RQ-PCR is the lability of RNA and the possibility of its application only in a small group of patients because of the limited frequency of gene fusions in ALL.

In the monitoring of MRD may be used the following gene fusions: *BCR-ABL1*, *MLL-AFF1*, *TCF3-PBX1* and *ETV6-RUNX1*, present in approximately 40% of pediatric and adults patients (Campana, 2009; Bruggemann et al., 2010), and *IGH@-CRLF2* or *P2RY8-CRLF2* abnormalities, which are detected in about 15% of adult or high risk pediatric B-ALL (Yoda et al., 2010).

The most common method of MRD detection is based on the sequence of rearranged Ig gene joints/TCR, encoding the immunoglobulin and T cell receptors in response real-time quantitative polymerase chain reaction. The method can be used in more than 95% of ALL patients and has a sensitivity of 10^{-4} to 10^{-5}. The Ig/TCR RQ-OCR method is complicated, time-consuming and cost-intensive, but nevertheless is regarded in Europe as the 'gold standard' in the monitoring of MRD. This is due to the large stage of standardization, comprising the steps research, primers and probes, as well as guidelines on the interpretation of the data obtained his allows receive high reproducibility of results in the determination of risk groups and assessment of MRD that is necessary in conducting multicenter studies (Witt et al., 2009).

3.3 Microarray analysis of genetic abnormalities in ALL

Microarray platforms for analysis of genetic alterations include cDNA array (detected large alterations, often over 100kB) and oligonucleotide arrays (oligo array, used short 20-50kB nucleotide probes). The potential of high-density microarray of specific cDNA sequences allows for hybridization of fluorescently labelled mRNA of leukemic cells. The value of this technique for the diagnosis and follow-up of ALL specific genetic abnormalities is enormous. Protein microarrays were used to investigate Notch-transduced signals in the development of T-cell ALL (Chan et al., 2007). The identification of novel miR genes expressed in different types of ALL forms the basis for further studies of the biology of ALL (Schotte et al., 2011).

4. Genetic differences and similarities amongst B and T derived ALL

4.1 B cell- acute lymphoblastic leukemia

During the B-cell ontogenesis the V,D and J segments are rearranged to generate a unique gene sequence for each cell. The distinctive genetic abnormalities occurring in B-derived ALL are as follows: clonal DJ rearrangements of *IGH@*gene, T-cell receptor rearrangements noticed in 70% of B-ALL, which are not helpful in differentiating from T-cell ALL, t(9;22) mainly found in adults patients, t(12;21) and hyperdiploidy (usually without structural abnormalities) occurring mainly in children. In B-type of ALL del(6q), del(9p), del(12p), t(17;19) and intrachrmomosomal amplification of chromosome 21 (iAMP21) are often detected.

The t(12;21) occurs most frequently in children, and probably arose early in pregnancy. This genetic abnormality causes the fusion of two genes AML1 and TAL, resulting disorder in an early stage of B cell development.

Adults most often found in t(9;22), which causes the fusion of BCR and ABL genes and leads to the fusion protein BCR-ABL1 (tyrosine-kinase) that interacts with multiple signal paths (eg RAS).

Translocation	Gen	Function	Prevalence % adults/children	Detection for MRD monitoring
t(9;22)(q34;q11)	BCR-ABL	Enhanced tyrosine kinase activity, which function in intracellular signalling pathways	<25>/5-8	mRNA
t(4;11)(q21;q23)	MLL-AF4	Transcription factor in the regulation of differentiation pathways	6/2-7 mainly in infants	mRNA
t(1;19)(q23;p13)	E2A-PBX	Transcription factor	/<6 (25-30% in pre-B-ALL)	mRNA
t(12;21)(p13;q22)	TEL-AML1	Transcription factor	<2/1-2	mRNA
11q23 aberrations	MLL and any one of fusion partners	Transcription factor	3-4/5-6	mRNA

Table 3. Structural chromosome abnormalities in B type of ALL

4.2 T cell- acute lymphoblastic leukemia

Clonal rearrangements of T-cell receptor (TCR) genes, an abnormal karyotype, translocations and chromosomal deletions almost always occurred in T-derived ALL (Table 4). About 20 % of patients diagnosed as T-ALL displayed immunoglobulin gene rearrangements as well (Szczepański et al., 1999). The abnormal karyotype is present in about 50-70% of cases and mainly involves the alpha and delta TCR(14q11.2), the beta locus (7q35) and the gamma locus (7p14-15).

Translocations

These genetic abnormalities are surrounded by translocations of the partner gene. The most frequently involved genes are as follows: HOX11 (TLX1) (10q24) occurred in 30% of adults and in 7% of pediatric patients, HOX11L2 (TLX3) (5q35) found in 10-15% of adults and in

20% of children. There are transcription factors. The other transcription factors engaged in translocations are *MYC* (8q24.1), *TAL1* (p32), *RBTN1* (*LMO1*) (11p15), *RBTN2* (*LMO2*) (11p13), *LyL1* (19p13) and the cytoplasmic tyrosine kinase *LCK* (1p34.3-35). In about 10-8% of patients other translocations occurred as follows: *PICALM-MLLT10* [*CALM-AF10*; t(10;11)(p13q14)] and *MLL* most often with the partner gene *ENL* (19p13). The translocations are often not detected by conventional cytogenetic methods hence PCR must be used (WHO, 2008).

Deletions

Del(9p) is the most frequently occurring deletion (in about 9% of cases), detected mainly by PCR method, only 30% can be detected by conventional cytogenetics (Brett-Gardiner et al., 2011).

Gene mutations

In over 50% of T-ALL cases *NOTCH1* gene mutations were found. The *NOTCH1* signalling pathway has three components and the mature *NOTCH1* protein which is essential for early T-cell development (Liu et al. 2011; Palomero & Ferrando A., 2009).

Brain and Acute Leukemia, Cytoplasmic (*BAALC*) gene expression

BALLC gene is located on chromosome 8q22.3. Its high expression (overexpression) in the T-cell ALL is associated with worse overall survival and relapse-free survival. Kuhn et al. demonstrated, that high *BAALC* expression is associated with inferior overall survival also in adults B-precursor ALL patients. *BAALC* overexpression can be regarded as an additional negative prognostic factor in adult ALL patients (Kuhnl et al., 2010).

5. Treatment of acute lymphoblastic leukemia patients

The main treatment rule for patients with acute lymphoblastic leukemia is to adjust the intensity of treatment to the level of disease aggression. This fact accounts for a wide application of protocols for risk-adapted therapy in everyday clinical practice. Clinical and biologic features which were defined as risk factors formerly (Table 5) are now replaced by minimal residual disease estimation (Attarbaschi et al., 2008; Bassan et al., 2009; Conter et al., 2010).

There are the same treatment protocols for B and T derived 'de novo' acute lymphoblastic leukemia. However, during the relapse of T-ALL other medications are recommended such as: Nelarabine, Forodesine and Clofarabine (De Angelo, 2009). Most European Groups use regimens containing prednisolone/dexamethasone, vincristine, daunorubicin and asparaginase in the induction phase of treatment of adult All patients (Conter et al., 2010; Bassan et al., 2009; Patel et al., 2010).

During consolidations additional chemotherapy using cyclophosphamide, cytarabine is administered, and including intensive intrathecal chemotherapy. Maintenance chemotherapy lasting 2-3 years consists of low-dose antineoplastic drugs, mainly of mercaptopurine and methotrexate (Holowiecki et al., 2006).

Hyper CVAD (cyclophosphamide, vincristine, adramycin and dexamethasone, without L-asparaginase) is a reasonable alternative for induction therapy and gives results similar to

Translocation	Gen	Function	Prevalence % adults/children	Detection for MRD monitoring
t(1;14)(p32;q11)	TAL1(SCL)	Transcription factor	/3	DNA (TAL1-TCRD)
t(11;14)(p15;q11)	RBTN1(i1)	Transcription factor	/9	DNA (LMO1-TCRD)
t(11;14)(p13;q11)	RBTN2(LMO2)	Transcription factor	/4-5	DNA (LMO2-TCRD)
t(10;14)(q24;q11) t(7;10)(q35;q24)	HOX11 HOX11	Transcription factor	/10	DNA (HOX11-TCRD or TCRB)
inv(7)(p15q34)	HOXA	Transcription factor	/5	mRNA (HOXA-TCRB)
t(5;14)(q35;q32)	HOX11	Transcription factor	10-15/20	mRNA (HOX11L2-BCL11B)
t(1;14)(p34;q11)	LCK	Transcription factor	3/3	DNA (TAL1-TCRD)
t(11;19)(q23;p13)	MLL-ENL	Transcription factor	/5	mRNA
t(7;9)(q34;q32) t(7;9)(q34;q34)	TAL2 NOTCH1	Transcription factor	~30/~50	mRNA
Del(9)(p21)	CDKN2A,CDKN2B	Loss of control cell cycle	/65-80	DNA/mRNA
del(1)(p32)	SIL-TAL1	Transcription factor	/10-25	DNA/mRNA (SIL-TAL1)
t(10;11)(p13-14;q14-21)	CALM	Transcription factor	10/8	mRNA (CALM-AF10)
MLL	MLL-ENL/AF10 etc	Transcription factor	8/5	mRNA (MLL-ENL/AF10 and other)
9p34 episomal amplification	NUP214-ABL	Enhanced tyrosine kinase activity, which function in intracellular signalling	/6	mRNA
inv(7)(p15q34)	HOXA	Transcription factor		mRNA

Table 4. Structural chromosome abnormalities in T type of ALL

	High risk factors (Hoelzer et al., 1988)	High risk factors PALG (Holowiecki et al., 2006)
Age	>35 years	≥ 35 years
WBC count	>30x10e9/l for B-type ALL >100x10e9 for T-type ALL	≥ 30x10e9/l for B-type ALL ≥ 100x10e9/l for T-type ALL
Immunophenotype	Prepre-B, early T, mature T	Prepre-B, early T, mature T
Genetics		t(4;11) or t(9;22)
Time to remission	CR after >4 weeks	MRD positivity post induction or during or post consolidation treatment

Table 5. Clinical and immunological prognostic factors for newly diagnosed adult ALL patients

the traditional induction protocols (Kantarijan et al., 2004). Various protocols for induction and consolidation as well as post remission treatment refer to pediatric patients and adults. In adult group of patients aged below 30 years some collaborative groups preferred more intensive treatment compatibility to pediatric protocols. This more intensive treatment gives better results defined as better complete remission rate (e.g. 83 vs 94% in the study of LALA-94 (adult) and FRALLE-93 (pediatric), respectively), disease free survival (e.g. 5 year EFS amounting 38% for adult (HOVON) and 71% for DCOG (pediatric) study, respectively), or overall survival (e.g. 5 year OS amounting respectively 56% (UKALLXII/E2993, adult) and 71% (ALL97, pediatric) (DeAngelo et al., 2007; Barry E. et al., 2007; Boissel et al., 2003; De Bont et al., 2005; Ramanujachar et al., 2007; Rijneveld et al., 2009; Testi et al., 2004).

For more advanced aged patients reduced intensity chemotherapy protocols are used generally due to lower tolerance to very intensive therapy (Giebel et al., 2010; Marks, 2010). In this age group significantly more frequently occurs high risk factors such as complex cytogenetic abnormalities, low hypodiploidy, t(4;11) and t(9;22). Philadelphia chromosome positivity (t(9;22)) is a very high risk factor but also allows the conduct of targeted therapy with the use tyrosine kinase inhibitors. In these cases, the cytogenetics methods allow the use of targeted therapy (Foa et al., 2011).

ALL occurs in children about five times more frequently than in adults. In recent years significantly increased the effectiveness of ALL treatment in the pediatric group of patients (Freyer et al., 2011). Surveillance, Epidemiology and End Results register (SEER) analysis showed the biggest improvement in survival of patients aged 15-19 years because the 5-year overall survival increased from 41 to 62%. Long term survival in between the ages 2 and 10 years are found in more than 90% (Stock, 2010). These good results of treatment may be due inter alia to the fact that children are usually treated within the prospective multicenter clinical trials, which included the optimal diagnostic methods and used most effective treatment.

Although many genetic abnormalities are important as prognostic factors, only a few have an influence on the choice of treatment (Tomizawa et al., 2007). There are some rules allowing for the introduction of certain particulars from diagnosis to modifications of

High risk factor	
WBC count	≥20G/l
Age	Infants or children ≥10 years
7-day response to steroid pretreatment	Presence of >1 x 10^9/l blasts in peripheral blood
Ploidy	< 45 chromosomes
Translocations:	t(9;22)/*BCR/ABL* t(4;11)/*MLL/AF4*
MRD after induction treatment	≥10^{-3}
CR after induction of treatment at the expected time (33 day of treatment)	No

Table 6. Adverse prognostic factors in acute lymphoblastic leukemia in children, which are most commonly used in everyday clinical practice.

patients' treatment and the results of cytogenetic findings are found to be useful. The first paper claiming cytogenetic abnormalities to have given important prognostic information was published by Secker-Walker (Secker-Walker et al., 1978). The impact of cytogenetics on the treatment results was published by Southwest Oncology Group 9400 study (Pullarkat et al., 2008). The authors presented the four different karyotype categories (Table 7) in which the most important prognostic factor for overall survival and relapse-free survival is cytogenetics instead of age.

Risk group	Cytogenetics markers
I. Standard risk (5 years overall survival ≥50%)	Hyperdiploid (>50 and < 66 chromosomes)
II. Intermediate risk (5 years overall survival 40-50%)	Normal diploid, 11q23 deletions without *MLL* rearrangements del9(p), del6(q), del(17p), del(12p), 13/del(13q), t(14q32),t(10;14) low hypodiploidy (47-50 chromosomes) *TCR* translocations Tetraploidy (>80 chromosomes)
III. High risk (5 years overall survival 30-40%)	t(1;19), 7(Ph-), del(7p), +8, 11q23/*MLL* gene and any one of fusion partners t(17;19), t(5;14)/*TLX3*, *CALM-AF*10
IV. Very high risk (5 years overall survival ≤30%)	t(9;22) *BCR-ABL*, t(4;11) *AF4-MLL*, t(8;14) *MYC-IGH* Complex (≥5 abnormalities without known translocations) Low hypodiloidy (30-39) Triploidy (60-78)

Table 7. Risk groups defined using karyotype and genetic categories based on SWOG analysis (Pullarkat et al., 2008)

Philadelphia chromosome positivity, low hypodiploidy (near triploidy) and complex cytogenetic abnormalities (more than five chromosomal changes) influenced shorter overall survival (Marks et al., 2009). Sometimes more intensive treatment e.g. modified Hyper-CVAD regimen gives better results in very high risk ALL with t(4;11) (Li et.al., 2009).

The translocation between chromosome 9 and 22 results in the formation of the Philadelphia chromosome (Ph) and generates the expression of a p 190 protein or encoding a chimaeric p210 protein. Ph positivity is more often present in adult patients and the incidence increases with the age from 20% in 30 years to 39% in over 60 years (Moorman A.V., 2010). In these cases, targeted therapy using tyrosine kinase inhibitors such as imatinib or dasatinib is combined with chemotherapy both in young and elderly patients (Foa et al., 2011; Ravandi et al., 2010; Laport et al., 2008; Ottmann et al., 2007; Tanguy-Schmidt et al., 2009). These types of combined therapy give better results mostly if allogeneic bone marrow transplantation was performed in the first complete remission (Yanada et al., 2008). In patients aged above 60 years chemotherapy should be reduced, and limited to administer monotherapy, imatinib 600-800mg/day, or in combination with glucocorticoids and/or vincristin (Ottmann, Wassmann et al., 2007).

5.1 Minimal residual disease testing

Many publications have shown that the MRD positivity has prognostic value for the treatment results of children and adults (Moricke et al., 2008; Stow et al., 2010; Hoelzer et al., 1988; Holowiecki et al., 2008). Result of the MRD examination is important in monitoring the induction and consolidation treatment (Figure 4), and in terms of eligibility for bone marrow transplantation (Couston-Smith et al., 2011; Giebel et al., 2010). MRD testing can be performed using the flow cytometry method or PCR and/or FISH.

In our paper published by Giebel et al. (Giebel et al., 2009) we documented that MRD measured by flow cytometry combined with cytogenetics replaces conventional risk criteria in adults with Ph-negative acute lymphoblastic leukaemia as the relapse rate is significantly lower in standard risk karyotype or intermediate risk karyotype and MRD lower than 0,1% after induction treatment according to the PALG 4-2002 protocol (Figure 5).

6. Allogeneic transplantation in ALL patients

Provide for the implementation of EBMT recommendations, bone marrow transplantation (the family donor and from an unrelated donor, and autotransplantation) during the second remission. For the decision on the application of allogeneic bone marrow transplantation (BMT) is to monitor MRD. Many research groups recommends allogeneic BMT in first remission period already, in patients at a high risk of leukemia, which is defined of t(9;22), t(4;11) occurrence and also hipodiploid (near haploid) karyotype (Ferra et al., 2010; Marks et al., 2008; Marks et al., 2010).

The presence of MRD after induction and/or consolidation treatment is also classified as high risk and is an indication for early performing allogeneic bone marrow transplantation. The sensitivity of MRD at a level of 10^{-5} can justify reducing the strength of the treatment in some specific cases. Patients of the latter group needs to step up therapy and eligibility for allogeneic bone marrow transplantation.

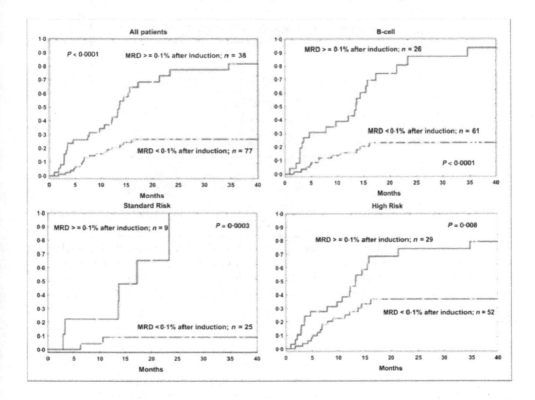

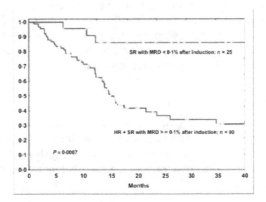

Fig. 4. The influence MRD test result on the effectiveness of treatment of adult ALL patients according to the protocol PALG 4-2002 (Holowiecki et al., 2008).

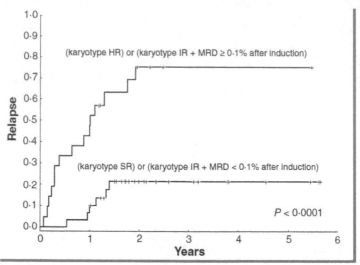

Fig. 5. Relapse incidence for Ph-negative ALL according to stratification criteria based on the combination of karyotype and minimal residual disease status. Karyotype risk groups were defined as proposed by Moorman et al (2007) SR indicates standard risk i.e. the presence of either del(9p) or high hyperdiploidy; HR, high risk i.e. the presence of t(4;11), t(8;14), hypodiploidy/almost triploidy or complex karyotype; IR, intermediate risk i.e. all remaining patients (Giebel et al., 2009).

7. Conclusions

The introduction of the WHO 2008 classification forces implementation of cytogenetic methods into the diagnostic procedures. Conventional cytogenetic and genetic aberrations analysis must be conducted in each case of acute lymphoblastic leukemia. Some genetic abnormalities cause the application of targeted therapy, the main example is the use of tyrosine kinase inhibitors in cases of the t(9;22). Cytogenetic methods, evaluating the minimal residual disease, are useful for optimization of the treatment strategy especially for recommendation of the allogeneic hematopoietic stem cell transplantation which is performed in high risk patients during the first complete remission.

8. Acknowledgment

The author wishes to thank colleagues of Clinic of Hematology and Bone Marrow Transplantation, Medical University of Silesia in particular to Professor Sławomira Kyrcz-Krzemień, the head of the Clinic for cooperation, and to Krystyna Jagoda and Aleksandra Bartkowska-Chrobak for the take advantage of the results laboratory investigations (Figure 1 and Figure2 and 3 succesively).

9. References

Anino L., Goekbuget N., Dellanoy A. (2002). Acute lymphoblastic leukemia in the elderly. *Hematology Journal*. Vol.3, No. 5, (September 2002), pp. 219-223, ISSN 1024-5332.

Apperly et al., Haematopoietic Stem cell Transplantation. The EBMT Handbook. (2008).European School of Haematology Publishing. Paris Cedex, France.

Attarbaschi A. et.al. (2008). Minimal residual disease values discriminate between low and high relapse risk in children with B-cell precursor acute lymphoblastic leukemia and an intrachromosomal amplification of chromosome21: The Austrian and German Acute Lymphoblastic Leukemia Berlin-Frankfurt-Munster (ALL-BFM) trials. *Journal of Clinical Oncology*. Vol.26, No.18, (June 2008), pp. 3046-3050, ISSN 0732-183X.

Balduzzi A. et al. (2011). Autologous purified peripheral blood stem cell transplantation compare to chemotherapy in childhood acute lymphoblastic leukemia after low-risk relapse. *Pediatric Blood & Cancer*. Vol. 57, No.4, (October 2011), pp. 654-659, ISSN 1545-5017.

Barry E. et al. (2007). Favorable outcome for adolescents with acute lymphoblastic leukemia treated on Dana-Farber Cancer Institute Acute Lymphoblastic Leukemia Consortium Prtotocols. *Journal of Clinical Oncology*. Vol.25, No.7, (March 2007), pp. 813-819, ISSN 0732-183X.

Bassan R. et al. (2009). Improved risk classification for risk-specific therapy based on the molecular study of minimal residual disease (MRD) in adult acute lymphoblastic leukemia (ALL). *Blood*. Vol.113, No18, (April 2009), pp. 4153-4162, ISSN 0006-4971.

Bene MC, (1995). Proposal for the immunological classification of acute leukemias. *Leukemia*. Vol. 9. (1995), pp. 1783-1786, ISSN 0887-6924.

Boissel N. et al. (2003). Should adolescents with acute lymphoblastic leukemia be treated as old children or young adults ? Comparison of the French FRALLE-93 and LALA-94 trials. *Journal of Clinical Oncology*. Vol.21, No 5, (March 2003), pp. 774-780, ISSN 0732-183X.

De Bont J.M. et al. (2005). Adolescents with acute lymphatic leukaemia achieve significantly better results when treated following Dutch paediatric oncology protocols than with adult protocols. *Nederlands Tijdschrift voor Geneeskd,* Vol. 149, No 8, (February 2005), pp. 400-406, ISSN 0028-2162.

Brem SS et.al. (2008). Central nervous system cancers. *Journal of National Comprehensive Cancer Network*. Vol.6, No, (2008), pp. 456-504, ISSN 1540-1405.

Brett Gardiner R et al. (2011). Using MS-MLPA as an efficient screening tool for detecting 9p21 abnormalities in pediatric acute lymphoblastic leukemia. *Pediatric Blood&Cancer*. Published online 27 Jul 2011 DOI:10.1002/pbc.23285 ISSN 1545-5017.

Bricarelli F.D. et al. (2006). Cytogenetic guidelines and quality assurance : A common European framework for quality asseeement for constitutional and acquired cytogenetic investigation. *European Cytegenetics Association Newsletter*. Vol. 17, No 17, (January 2006), pp. 13-32.

Broadfield Z.J., et al. (2004). Complex chromosomal abnormalities in utero, 5 years before leukemia. *British Journal of Haematology*. Vol.126, No. 3, (August 2004), pp. 307-312, Online ISSN: 1365-2141

Bruggemann M. et al., (2010). Standardized MRD quantification in European ALL trials: proccedings of the Second International Symposium on MRD assessment in Kiel, Germany;2008. *Leukemia*. Vol. 24, (September, 2010), pp. 521-535, ISSN 0887-6924.

Campana D. (2009). Minimal residual disease in acute lymphoblastic leukemia. *Seminars in Hematology*. Vol. 46, No. 1, (January 2009), pp. 100-106, ISSN 0037-1963.

Chang P. et al. (2010). FLT3 mutation incidence and timing of origin in a population case series of pediatric leukemia. *BMC Cancer*. 10;513; (2010), doi:10.1186/1471-2407-10-513.

Chan S. et al. (2007). Notch signals positively regulate activity of the mTOR pathway in T-cell acute lymphoblastic leukemia. *Blood*, Vol. 110, No. 1, (July 2007), pp. 278-286, ISSN 0006-4971.

Conter V. et al. (2010). Molecular response to treatment redefines all prognostic factors in children and adolescent with B-cell precursor acute lymphoblastic leukemia: results in 3184 patients of the AIEOP-BFM ALL 2000 study. *Blood*. Vol.115, No.16, (April 2010), pp. 3206-3214, ISSN 0006-4971.

Coustan-Smith E et al. (2011). New markers for minimal residual disease detection in acute lymphoblastic leukemia. *Blood* Vol. 117, No. 23, (June 2011), pp 6267-6276; ISSN: 0006-4971.

DeAngelo DJ. Et al., (2007). A multicenter phase ii study using a dose intensified pediatric regimen in adults with untreated acute lymphoblastic leukemia (Abstract). *Blood*, Vol. 110, (2007), p. 587, ISSN 0006-4971.

De Angelo. (2009). Nelarabine for the treatment of patients with relapsed or refractory T-cell acute lymphoblastic leukemia or lymphoblastic lymphoma. *Hematology/Oncology Clinics of North America*. Vol. 23, No. 5, (2009), pp. 1121-1135, ISSN 0889-8588.

Delgado M.M. et al. (2010). Myc roles in hematopoiesis and leukemia. *Genes Cancer*. Vol.1, No.6, (June 2010), pp. 605-616, ISSN 1947-6019.

Faderl S. et al. (1998). Clinical significance of cytogenetic abnormalities in adult acute lymphoblastic leukemia. *Blood*, Vol. 91, No. 11, (November 1998), pp. 3995-4019, ISSN 0006-4971.

Faderl S. et al. (2010). Adult acute lymphoblastic leukemia: concepts and strategies. *Cancer*. Vol.116, No. 5, (March 2010), pp. 1165-1176, ISSN 1097-0142.

Ferra Ch. et al. (2010). Unrelated transplantation for poor-prognosis adult acute lymphoblastic leukemia: Long-term outcome analysis and study of the impact of hematopoietic graft source. *Biology of Blood and Marrow Transplantation*. Vol.16, No. 7, (July 2010), pp. 957-966, ISSN 0268-3369.

Foa R. et al. (2011). Dasatinib as first-line treatment for adult patients with Philadelphia chromosome-positive acute lymphoblastic leukemia. *Blood*, doi: 10.1182/blood-2011-05-351403 (published online before print)

Freyer D.R et al. (2011). Postrelapse survival in childhood acute lymphoblastic leukemia is independent of initial treatment intensity: a report from Children's Oncology Group. *Blood*. Vol. 117, No. 11, (March 2011), pp. 3010-3015, ISSN 0006-4971.

Giebel S. et al.(2009). Could cytogenetics and minimal residual disease replace conventional risk criteria in adults with Ph-negative acute lymphoblastic leukaemia? *British Journal of Haematology*. Vol. 144, No. 6, (March 2009), pp. 970-972, Online ISSN 1365-2141.

Giebel S. (2010). Treatment of acute lymphoblastic leukemia in elderly patients. *Hematologia*. Vol.1, No. 1, (January 2010), pp. 41-48, ISSN 2081-0768.

Giebel S. et al. (2010). Status of minimal residual disease determines outcome of autologous hematopoietic SCT in adult ALL. *Bone Marrow Transplantation*. Vol.45, No.6, (June 2010), pp. 1095-1101, ISSN 0268-3369.

Haferlach C. et al. (2007). Proposals for standardized protocols for cytogenetic analyses of acute leukemias, chronic lymphocytic leukemia, chronic myeloid leukemia, chronic myeloproliﬁetative disorders, and myelodysplastic syndromes. *Genes, Chromosomes and Cancer*. Vol. 46, No. 5, (May 2007), pp. 494-499, ISSN 1098-2264.

Harrison Ch. J. et al. (1999). The value of multiple colour FISH in the cytogenetic analysis of leukaemia. *Cytogenetics and Cell Genetics*. Vol. 85, (1999), pp. 745, ISSN 0301-0171.

Harrison Ch.J. (2001). Acute lymphoblastic leukaemia. *Best Prectice & Research Clinical Haematology*, Vol.14, No.3, (2001), pp. 593-607, ISSN 1521-6926.

Harrison Ch.J. (2008). Cytogenetics of paediatric and adolescent acute lymphoblastic leukaemia. *British Journal of Haematology*, Vol. 144, No. 2, (November 2008), pp. 147-156, Online ISSN 1365-2141.

Hoelzer D. et al. (1988). Prognostic factors in a multicenter study for treatment of acute lymphoblastic leukemia in adults. *Blood*. Vol.71, No. 1, (January 1988), pp. 123-131, ISSN 0006-4971.

Hoelzer D., Gokbuget N. (2009). T-cell lymphoblastic lymphoma and T-cell acute lymphoblastic leukemia: a separate entity?. *Clinical Lymphoma and Myeloma*, Vol.9, Suppl. (3, 2009), pp. 214-21, PMID 19778844; ISSN 1557-9190.

Holowiecki J. et al. (2006). Minimal residual disease status is the most important predictive factor in adults with acute lymphoblastic leukemia. PALG 4-2002 prospective MRD study. *Hematologica/The Hematology Journal*. Vol.91, Suppl. 1, (2006), p. 360, ISSN 1466-4860.

Holowiecki J.et al. (2008). Status of minimal residual disease after induction predicts outcome in both standard risk and high risk Ph-negative adult lymphoblastic leukemia. The Polish Adult Leukemia Group ALL 4-2002 MRD Study. *British Journal of Haematology*. Vol. 142, No.2, (July 2008), pp. 227-2137, Online ISSN 1365-2141.

ISCN (2005). An international system for human cytogenetic nomenclature. Karger, Basel, 2005, ISBN 3-8055-8019-3

Kallioniemi A. et al. (1994). Optimizing comperative genomic hybridization for analysis of DNA sequences copy number changes in solid tumors. *Genes, Chromosomes and Cancer*. Vol.10, No.4, (August 1994), pp. 231-243, ISSN1098-2264.

Kantarjian H. et al. (2004). Long-term follow-up results of hyperfractionated cyclophosphamide, vincristine, doxorubicin, and dexamethasone (Hyper-CVAD), a dose intensive regimen, in adult acute lymphocytic leukemia. *Cancer*. Vol.101, No. 12, (December 2004), pp. 2788-2801, ISSN 1097-0142.

Kowalczyk et al. (2010). Structural and numerical abnormalities resolved in one-step analysis: the most common chromosomal rearrangements detected by comparative genomic hybrydization in childhood acute lymphoblastic leukemia. *Cancer Genetics and Cytogenetics*. Vol. 200, No. 2, (July 2010), pp. 161-166, ISSN 0165-4608.

Kuhnl A. et.al (2010). High BAALC expression predicts chemoresistance in adult B-precursos acute lymphoblastic leukemia. *Blood*. Vol.115, No.18, (May 2010), pp. 3737-3744, ISSN 0006-4971.

Laport G et.al. (2008). Long-term remission of Philadelphia chromosome positive acute lymphoblastic leukemia after allogeneic hematopoietic cell transplantation from matched sibling donors: a 20 year experience with the fractionated total body irradiation-etoposide regimen. *Blood*, Vol.112, No. 3, (August 2008), pp. 903-909, ISSN 0006-4971.

Li Y. et.al. (2009). Clinical characteristics and treatment outcome of adult acute lymphoblastic leukemia with t(4;11)(q21q23) using a modified Hyper-CVAD regimen. *Acta Haematologica*. Vol.122, No. 1, (October 2009), pp. 23-26, ISSN 0001-5792.

Liu H. et al. (2011). Critical roles of NOTCH1 in acute T-cell lymphoblastic leukemia. *International Journal of Hematology* DOI 10.1007/s12185-011-0899-3; ISSN: 0925-5710.

Marks D.I. et al. (2008). Unrelated donor transplants in adults with Philadelphia-negative acute lymphoblastic leukemia in first complete remission. *Blood*. Vol.112, No.2, (July 2008), pp. 426-434, ISSN 0006-4971.

Marks D.I. et al. (2009). T-cell acute lymphoblastic leukemia in adults : clinical features, immunophenotype, cytogenetics, and outcome from the large randomized prospective trial (UKALL XII/ECOG 2993) *Blood*. Vol.114, No. 25, (December 2009), pp. 5136-5145, ISSN 0006-4971.

Marks D.I. (2010). Treating the 'older' adult with acute lymphoblastic leukemia. *Hematology 2010. American Society of Hematology Education Program Book.* Orlando, Florida pp. 13-20.

Marks D.I. et al. (2010). The outcome of full-intensity and reduced-intensity conditioning matched sibling or unrelated donor transplantation in adults with Philadelphia chromosome-negative acute lymphoblastic leukemia in first and second complete remission. *Blood*. Vol.116, No.3, (July 2010), pp. 366-374, ISSN 0006-4971.

McGrattan P. et al. (2008). Integration of conventional cytogenetics, comparative genomic hybridisation and interphase fluorescence in situ hybridization for the detection of genomic rearrangements in acute leukaemia. *Journal of Clinical Pathology*. Vol.61, No.6, (August 2008), pp. 903-908, ISSN 0021-9746.

Moorman A.V. et al., (2010). A population-based cytogenetic study of adults with acute lymphoblastic leukemia. *Blood*. Vol. 115, No.2, (January 2010), pp. 206-214, ISSN 0006-4971.

Moricke A. et al. (2008). Risk-adjusted therapy of acute lymphoblastic leukemia can decrease treatment burden and improve survival: treatment results of 2169 unselected pediatric and adolescent patients enrolled in the trial ALL-BFM 95. *Blood*. Vol.111, No. 9, (May 2008), pp. 4477-4489, ISSN 0006-4971.

Mullighan C.G.(2009). Genomic profiling of acute lymphoblastic leukemia. *Hematology Education: the education program for the annual congress of the EHA*. Vol.3, No. 1, (June 2009), pp. 1-7.

Ness G.O. et al. (2002) Usefulness of high-resolution comparative genomic hybridization (CGH) for detecting and characterizing constitutional chromosome abnormalities. *American Journal of Medical Genetics*. Vol.113, No.2, (November 2002), pp. 125-136, ISSN 1552-4833.

Nijmeijer B.A.et al.(2010). A mechanistic rationale for combining alemtuzumab and rituximab in the treatment of ALL. *Blood*. Vol.116, No. 26, (December 2010), pp. 5930-5940, ISSN: 0006-4971.

Ottmann O.G. Wassmann et al. (2007). Imatinib compared with chemotherapy as front-line treatment of elderly patients with Philadelphia chrmomosome-positive acute lymphoblastic leukemia (Ph+ALL). *Cancer*. Vol. 109, No.10, (May 2007), pp. 2068-2076, ISSN 1097-0142.

Ottmann O.G. et al. (2007). Dasatinib induces rapid hematologic and cytogenetic responses in adult patients with Philadelphia chromosome-positive acute lymphoblastic leukemia with resistance or intolerance to imatinib: interim results of a phase 2 study. *Blood*, Vol.110, No. 7, (October 2007), pp. 2309-2315, ISSN 0006-4971.

Palomero T. & Ferrando A. (2009) Therapeutic targeting of NOTCH1 signaling in T-ALL. *Clinical Lymphoma and Myeloma*. Vol.9, (Suppl 3, 2009) pp. 205-210, ISSN 1557-9190

Patel B. et al., (2010). Minimal residual disease is a significant predictor of treatment failure in non T-lineage adult acute lymphoblastic leukemia: final results of the international trial UKALL XII/ECOG2993. *British Journal of Haematology*, Vol. 148, No.1, (January 2010) pp. 80-89, Online ISSN: 1365-2141.

Park JS., et al. (2008) Comparison of multiplex reverse transcription polymerase chain reaction and conventional cytogenetics as a diagnostic strategy for acute leukemia. *International Journal of Laboratory Hematology*. Vol.30, No. 6, (December 2008), pp. 513-518, ISSN: 1751-553X.

Pullarkat V et al. (2008). Impact of cytogenetics on the outcome of adult acute lymphoblastic leukemia: results of Southwest Oncology Group 9400 study. *Blood*, Vol. 111, No.5, (March 2008), pp.2563-2572, PMC2254550.

Raetz E.A. et al. (2006). Gene expression profiling reveals intrinsic differences between T-cell acute lymphoblastioc leukemia and T-cell lymphoblastic lymphoma. *Pediatric Blood and Cancer*. Vol. 47, No.2, (August 2006), pp. 130-140, ISSN 1545-5017.

Raetz E.A. et al (2008). Chemoimmunotherapy reinduction with Epratuzumab in children with acute lymphoblastic leukemia in marrow relapse: A Children's Oncology Group Pilot Study. *Journal of Clinical Oncology*, Vol.26, No. 22, (August 2008), pp. 3756-3762, ISSN: 0732-183X.

Ramanujachar R., et al (2007). Adolescents with acute lymphoblastic leukemia: outcome on UK national paediatric (ALL97) and adult (UKALLXII/E2993) trials. *Pediatric Blood and Cancer*, Vol. 48, No.3, (March 2007), pp. 254-261, ISSN 1545-5017.

Ravandi F. et al., (2010). First report of phase 2 study of dasatinib with hyper-CVAD for the frontline treatment of patients with Philadelphia chromosome-positive (Ph+) acute lymphoblastic leukemia. *Blood*. Vol. 116, No. 12, (September, 2010), pp. 2070-2077, ISSN 0006-4971.

Rhein P., et al. (2010). CD11b is a therapy resistance – and minimal disease-specific marke in precursor B-cell acute lymphoblastic leukemia. *Blood*, Vol.115, No. 18, (May 2010), pp. 3763-3771, ISSN 0006-4971.

Rijneveld A.W. et al. (2009). High dose intensive chemotherapy, as is standard in childhood leukemia, is feasible and efficacious in adult patients with acute lymphoblastic leukemia (ALL) up to the age 40: Results from the Dutch-Belgian HOVON-70 Study. *Blood*. (ASH Annual Meeting Abstracts, December 2009) Vol. 114, p.323, ISSN 0006-4971.

Schotte D. et al. (2011). Discovery of new microRNAs by small RNAome deep sequencing in childhood acute lymphoblastic leukemia. *Leukemia*. Vol.25, No. 9, (September 2011), pp. 1389-1399, ISSN 0887-6924.

Secker-Walker LM et al. (1978). Prognostic implications of chromosomal findings in acute lymphoblastic leukaemia at diagnosis. *British Medical Journal*, Vol.2(6151), No. 2, (December 1978), pp. 1529-1530, PMCID: PMC1608754; ISSN 0959-8138.

Soszynska K. et al. (2008). The application of conventional cytogenetics, FISH and RT-PCR to detect genetic changes in 70 children with ALL. *Annals of Hematology* Vol.87, No. 12, (December 2008), pp. 991-1002 ISSN 0939-5555.

Stock W., (2010). Adolescent and young adults with acute lymphoblastic leukemia. *Hematology 2010. American Society of Hematology Education Program Book*. Orlando, Florida, USA, pp. 21-29.

Stow P. et al. (2010). Clinical significance of low levels of minimal residual disease at the end of remission induction therapy in childhood acute lymphoblastic leukemia. *Blood*. Vol. 115, No. 23, (June 2010), pp. 4657-4663, ISSN 0006-4971.

Szczepanski T. et al. (1999). Ig heavy chain gene rearrangements in T-cell acute lymphoblastic leukemia exhibit predominant DH6-19 and DH7-27 gene usage, can result in complete V-D-J rearrangements, and are rare in T-cell receptor alpha beta lineage. *Blood.* Vol.93, No.12, (June 1999), pp. 4079-4085, ISSN 0006-4971.

Testi A.M. et al., (2004). Difference in outcome of adolescents with acute lymphoblastic leukemia reenrolled in pediatric (AE10P) and adult (GIMEMA) protocols. *Blood.* Vol. 104, No.2, (July 2004), 539, ISSN 0006-4971.

Tomizava D. et.al (2007). Outcome of risk-based therapy for infant acute lymphoblastic leukemia with or without an MLL gene rearrangement, with emphasis on late effects: a final report of two consecutive studies, MLL96 and MLL98, of the Japan Infant Leukemia Study Group. *Leukemia.* Vol.21, No. 11, (September 2007), pp. 2258-2263, ISSN 0887-6924.

Topp M.S. et al.(2009) Report of a phase II trial of single-agent BiTE® antibody Blinatumomab in patients with minimal residual disease (MRD) positive B-precursor acute lymphoblastic leukemia (ALL). *Blood* (ASH Annual Meeting Abstracts, December 2009). Vol. 114, p. 840, ISSN 0006-4971.

Topp M.S. et al. (2011). Targeted therapy with the T-cell-engaging antibody blinatumomab of chemotherapy-refractory minimal residual disease in B-lineage acute lymphoblastic leukemia patients results in high response rate and prolonged leukemia-free survival. *Journal of Clinical Oncology.* Vol. 29, No. 18, (June 2011), 2493-2498, ISSN 0732-183X.

Tanguy-Schmidt A. et al. (2009). Long-term results of the imatinib GRAAPH-2003 study in newly-diagnosed petients de novo Philadelphia chromosome-positive acute lymphoblastic leukemia. *Blood* (ASH Annual Meeting Abstracts, December 2009) Vol. 114, p. 3080, ISSN 0006-4971.

Wesołowska A et al.(2011). Cost-effective multiplexing before capture allows screening of 25000 clinically relevant SNPs in childhood acute lymphoblastic leukemia. *Leukemia.* Vol. 25, No. 6, (June 2011), pp. 1001-1006, ISSN 0887-6924.

Witt, M., Szczepański, T. & Dawidowska, M. (2009). *Hematologia molekularna, patogeneza, patomechanizmy i metody badawcze.* Ed: Ośrodek Wydawnictw Naukowych, Poznań, Polska.

Wood B.et al. (2006). 9-color and 10 color flow cytometry in the clinical laboratory. *Archives of Pathology and Laboratory Medicine.* Vol. 130, No. 5, (May 2006), pp. 680-690, ISSN 0003-9985.

WHO Classification of Tumours of Haematopoietic and Lymphoid Tissues. (2008), Edited by Steven H et al., ISBN 978-92-832-2431-0, Lyon, France.

Yanada M. et al. (2008). Karyotype at diagnosis is the major prognostic factor predicting relapse-free survival for patients with Philadelphia chromosome-positive acute lymphoblastic leukemia treated with imatinib-combined chemotherapy. *Haematologica.* Vol. 93, No.2, (February 2008), pp. 287-290, ISSN 0390-6078.

Yoda et al., (2010). Functional screening identifies CRLF2 in precursor B-cell acute lymphoblastic leukemia. *Proceedings of National Academy of Sciences, USA.* Vol. 107, (2010), pp. 252-257, ISSN 0027-8424.

7

Cytogenetic Instabilities in Atomic Bomb-Related Acute Myelocytic Leukemia Cells and in Hematopoietic Cells from Healthy Atomic Bomb Survivors

Kimio Tanaka

Department of Radiobiology, Institute for Environmental Science, Aomori
Japan

1. Introduction

On August 6 and 9, 1945, nearly 65 years ago, atomic bombs (ABs) were dropped on two Japanese cities, Hiroshima and Nagasaki. The explosive powers of these two bombs are said to be equal about 12.5 kilotons of TNT for Hiroshima bomb and 22 kilotons for Nagasaki bomb. The dissipation of energy is believed to have been in the ratios consisting of 50% of bomb blast, 35% of thermal rays and 15% of radiation. Heat and bomb blast caused death in AB survivors within two weeks after bombing. On the other hand cause of death after two weeks were closely related to radiation. The initial radiation caused by the Hiroshima bomb composed mostly gamma rays including about 10 % of neutrons, on the while Nagasaki bomb emitted only gamma rays, In the both cities, a 50% death rate was estimated for those who were exposed at 1.2 km on the hypocenter. LD50 value in human was estimated as around 4 Gy from Hiroshima AB victims. Bone marrow depletion by radiation was the most critical damage leading to death. There victims who died early had extensive bone marrow damage, manifested by leukopenia and thrombocytopenia. Epilation began one to four weeks post-exposure.

Epidemiological studies have established that leukemia develops more frequently among atomic bomb (AB) survivors than in the general populations(Pierce *et al.*1996). Observed cancers and the approximate dates at which significant increases became evident were leukemia in 1950, thyroid 1955, breast and lung cancer in 1965, and gastric and colon cancer in 1975, and multiple myeloma in 1975. The incidence of leukemia peaked 8-10 years after the bombing, and a higher incidence was found in survivors exposed to higher radiation doses and those exposed at a younger age. The incidences of chronic myelocytic leukemia (CML) and acute lymphocytic leukemia(ALL) were decreased to the control level after 15-20 years and 25-30 years after the bombing, respectively, while the relative risk of AML development is still increased in persons who were exposed to the bombing at the age of 20-39 years（Preston *et al.*1994). The peak leukemia prevalence among Hiroshima survivors has passed, but even in recent years the death rate from leukemia among proximately exposed survivors remains higher than the mean death rate of the non-exposed and for all Japan. Among those exposed within 2 km of the hypocenter of Hiroshima, there was a marked increase in CML and the ratio of acute type to chronic type was 1.6, whereas almost the same ratio as the all-Japan ratio of 3.8 was observed in those exposed beyond 2 km (Kamada and Tanaka 1983). The Leukemia Registry Study has revealed that the younger the individual at the time of bombing , the greater was the risk of leukemia during the early period and the more rapid was decline thereafter. On the other hand, in the group aged 45 or more at the time on bombing, the increase in risk occurred later and persisted during the period 1960-1975. Acute type leukemia contributed to these trends. More recent reports have indicated that the incidences of acute myelocytic leukemia (AML) and MDS were still high even 40-60 years after the bombing (Richardson *et al.*, 2009; Iwanaga *et al.*, 2011). These results suggest that mechanism for development of radiation-induced AML seems to be different those of ALL and CML.

In summary, AB-related leukemia and solid tumor exhibit the following characteristics: (1) Leukemia and cancer risk increased with dose. (2) The leukemia risk increased with decreasing age at the time of bombing. (3) Unlike leukemia, the latency period increase with decreasing age at the time bombing, with a marked radiation effect evident when survivors reached the age level at which the cancers frequently occur. AB-survivors population is very unique and can be used for clarifying the mechanisms responsible for radiation-induced leukemia.

2. Dosimetry system of AB radiation

The former T65D system of dosimtry had been used as the most accurate method of estimating the doses received by individual survivors although this has now been replaced by the DS86 and DS01 system. Original T65D system was based on the nuclear testing conducting at Nevada, USA, and was devised in 1965 as a formula that incorporated various parameters such as distance from the hypocenter, and transmission factors for shielding materials, etc. AB survivors are defined as those directly exposed in the city at the time of the bombing within 5 km of the hypocenter. In order to permit more detail calculations, the DS86 and DS01 system was formulated in 1986 and 2001 based on elementary physical processes, and enabled computer analysis of the different processes involved from the time of emission until arrival at various human organs, including factors of calculations of weather and soil. Dose of gamma rays was increased four times, on the other hand, neuron dose is decreased more than 10% of former values. The DS86 system was used for present study. The DS86 assessment of Hiroshima AB radiation contained low energy neutrons of less than 1.0 MeV. If the hypothesis that the RBE of low energy neutrons is higher than that of fission neutrons with 2.3 MeV is true, Hiroshima AB radiation would have created a significantly higher degree of biological effects. Therefore our experiments using ^{252}Cf-neutrons and monochromatic neutrons was performed in Hiroshima University, The relative biological effectiveness (RBE) of dicentric chromosomes aberrations was increased to 10.7 at 0.75 MeV from 3.9 at ^{252}Cf- neutrons and reached to 16.4 as highest RBE value at 0.37 MeV, but the value was decreased to 11.2 at 0.186 MeV (Tanaka *et al.* 1994, 1999). RBE of low energy neutron is higher than that of fission neutrons. It is still now in question how radiation containing low energy neutron effects on developments of leukemia and cancer.

3. Chromosome aberrations in AB-related leukemia

We have investigated chromosome aberrations in leukemias developing individuals with a history of heavily exposure to AB radiation(Kamada *et al.*,1991, Nakanishi *et al.*, 1999) and have reported cytogenetic and molecular biological alterations in AB-related leukemias (Kamada and Tanaka 1983; Tanaka *et al.*, 1989; Tanaka *et al.*, 1991; Kamada *et al.*, 1991). In the present study, to clarify the mechanisms responsible for radiation-induced leukemia, we analyzed cytogenetic and molecular biological alterations in AB-related leukemias in comparison with those in *de novo* (non-AB-related) leukemia. Acute lymphocytic leukemia (ALL), chronic myeloid leukemia (CML) and other hematological disorders were excluded from the present analysis. In our previous cytogenetic molecular biological studies of 132 patients with AB-related leukemia observed during 1978-1999, 33 patients with acute myeloid leukemia (AML) had been exposed to DS86 doses of exceeding 1Gy. Eight more AB patients were included in the present study, after our previous publication in 1999 (Nakanishi *et al.*, 1999). This revealed that among 132 patients with AB-related leukemias observed during 1978-1999, 33 patients with AML had been exposed to doses of more than 1 Gy on the basis of the Dose System 1986 (DS86). Leukemia patients were divided into three groups according to their exposure status; those who received more than 1 Gy bone marrow dose, 0.01-0.99 Gy and non-exposed groups (Table 1). Chromosome aberrations in the 33 patients were compared with those in 588 control patients with *de novo* AML who had been born before August 1945. Thirty-two and 58 AB-related patients who had been exposed to 0.01-0.99 Gy and less than 0.01Gy were also analyzed for chromosome aberrations using the same methods.

Bone marrow cells or peripheral lymphocytes from 33, 32, 58 and 588 leukemia patients with and without AB radiation exposure, respectively, were cultured for 24 h and harvested for chromosome preparations. These preparations were stained by the Giemsa banding procedure and karyotyped according to ISCN (2009). Mean numbers of aberrant chromosomes were scored in karyotypes of the main leukemia clone in each patient. One thousand metaphases were observed by microscopy, and spontaneous chromosome aberrations such as chromatid breaks, gaps, and hyperploid cells, which are useful markers of radiation-induced chromosome instability, were also detected. FISH analyses were performed using cosmid and YAC probes such as *MLL* and *CD3* on 11q22-23 regions of chromosome 11 for mapping of chromosome breakpoints in each patient.

The leukemia subtypes of FAB classification in the patients in each group were analyzed. Among 33 AB-related patients exposed more than 1Gy, 13 had been diagnosed as having MDS (RA or RAEB) by the FAB classification before the development of AML. The others had been diagnosed as having AML (AMLM1- 8 patients; AMLM2- 8 patients; AMLM4- 1 patient; AMLM5- 1 patient; AMLM6- 1 patient). None of the patient had AML M3.

Chromosome complexity was indicated by the number of aberrant chromosomes found in the main leukemic clone. Of the 33 AB-related patients who had been exposed to > 1Gy, only one (3.1%) had a normal karyotype, compared with 45.7% of AB-related patients exposed to 0.01-0.99 Gy, and 62.5% exposed to < 0.01 Gy. On the other hand, 44.1% of the non-exposed control leukemia patients (246 out of 558) had a normal karyotype. The numbers of aberrant chromosome per metaphase were 3.69, 1.89, 2.0 and 0.93 in AB-related leukemia patients who had been exposed to>1Gy, 0.01-0.99 Gy, <0.01Gy and non-exposed, respectively. These results indicate that AB-related patients who had been exposed >1.0 Gy had more complex chromosome aberrations in their leukemic cells. Both-AB-leukemia groups with exposure to 0.01-0.99 Gy and <0.01 Gy also had higher numbers of aberrant chromosomes per cell. Only 3.1% of patients with AB-leukemia who had been exposed to >

1 Gy had normal karyotype, while 45.7% and 62.5% of those exposed to 0.01-0.99 Gy and <0.01 Gy did so. About half of the AML patients showed normal karyotype.

Specific chromosome aberrations that were found at high frequencies are also listed in Table 1. Deletion of chromosome 20[del(20)] and loss of chromosome 20 were found in 10 patients, and der(11)or del(11) at 11q13-11q22 of chromosome 11 in 10 patients. On the other hand, non-exposed patients had a higher number of t(8;21) and t(15;17) translocations, which is specific to AML FAB type M2 and M3,respectively. The groups exposed to 0.01-0.99 Gy and <0.01Gy had no representative chromosome aberrations and lower percentages of t(8;21) and t(15;17) than the patient with *de novo* AMLs.

DS86 dose (Gy)	Number of patients	Percentage with a normal karyotype	Number of aberrant chromosomes per metaphase	Specific chromosome aberrations
AB patients exposed to >1.0Gy	33	3.1%	3.69	del(20)(q11) der(11)(q23) der(11)(q13)
0.01-0.99Gy	32	45.7%	1.89	-
<0.01Gy	58	62.5%	2.0	-
Non-exposed patients	558	44.1%	0.93	t(8;21) t(15;17)

Table 1. Comparison of chromosome aberrations found in patients with AB leukemia and patients with *de novo* leukemia

The incidences of several leukemia-specific chromosome aberrations were compared between the 33 patients with AB-related leukemia who were exposed to >1 Gy or non-exposed, and the 558 patients with *de novo* leukemias (Fig.1). Deletion of chromosome 5 or loss of chromosome [del(5)/-5] showed the highest incidence, but the incidences did not differ between the two groups. Deletion of chromosome 7 or monosomy of chromosome 7[del(7)/-7] showed the second highest incidence, and these aberrations were found more frequently in *de novo* leukemias. Deletion (20)/-20 and deletion and derivative of chromosome 11 at 11q13-11q22 were observed in higher frequencies in patients with AB-related leukemia. Deletion or monosomy of chromosome 13 [del(13)/-13] was also more frequent in AB-related leukemias [2 of 33 in del(13) and 3 of 33 in monosomy 13]. Translocations of t(8;21), t(15;17), inv(16) and t(9;22) were detected only in *de novo* leukemias. Further FISH analysis using target gene probes were performed, and the results indicated that the breakpoints associated with derivative translocations at 11q22-23 of chromosome 11 [der(11)t(11;α)(q22-23;α)]lay outside the *MLL* gene. FISH analysis using whole chromosome painting of chromosome 8 and target gene probes demonstrated severe complex chromosome aberrations containing chromosome segmental jumping translocations(SJT) at 8q24 of chromosome 8 and 11q22-23 of chromosome 11 (Tanaka *et al*, 2001). *MYC* oncogene signals located on some regions on the long arm of chromosome 8 and several other chromosomes, which were considered to translocate to several other chromosomes from original region on chromosome 8 (8q24). The phenomenon is known as segmental jumping chromosomal translocation(SJT) (Tanaka *et al.*, 1997). In conclusion, AB-

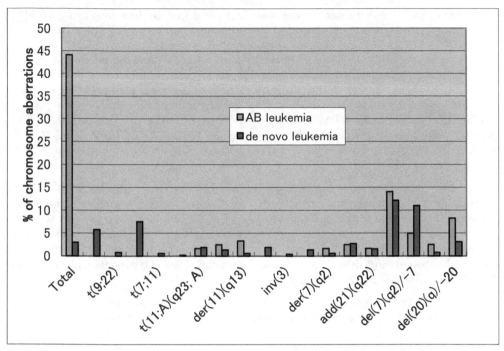

Fig. 1. Comparison of frequencies of each type of chromosome aberration between AB-
related leukemias and *de novo* leukemias

related leukemia patients showed more complex chromosome aberrations and did not have
any leukemia-specific chromosome aberrations such as t(8:21) and t(9;22).

4. Oncogene alterations in leukemias developed from atomic bomb survivors

Alterations in the *AML1* and *MLL* oncogenes in AML patients were also analyzed by PCR-
SSCP and FISH, respectively, and compared with those in the *de novo* in AMLs. Stored
DNAs were available from 9 patients with AB-related leukemia, including 6 patients who
had been exposed > 1.0 Gy. These leukemia DNAs were analyzed for *AML1* oncogene
alterations by PCR-SSCP. Exons 1 and 3 in the *AML1* gene, which overlaps the RUNT
domain of the *AML1* gene were analyzed.

Runt domain contains exons 3, 4 5, 6 and 7. PCR-SSCP and DNA sequence analyses of the
AML1 gene on human chromosome 21 were performed in 8 AML patients who had been
heavily exposed to the AB radiation. Genomic DNAs were amplified by PCR in a total
volume of 20 µl containing PCR buffer. PCR of exons 1 and 3 of the *AML1* gene was
performed using forward/reverse primers for the flanking intron. To identify *AML1* gene
mutations, PCR-SSCP analysis was performed on a Gene Phor system (Amersham
Pharmacia Biotech). After gelelectrophoresis, the gels were silver-stained to visualize the
bands. All PCR products with abnormal bands detected by PCR-SSCP were confirmed by
independent amplification and PCR-SSCP analysis. For identification of *AML1* gene
mutations, PCR products that showed abnormal bands were subcloned into a pCR2.2 vector

(Invitrogen), and 10 independent clones were sequenced in both directions using a BigDye Terminator Cycle sequencing kit (Perkin-Elmer) and analyzed on an ABI Prism 310 Genetic Analyzer (Perkin Elmer). To confirm mutations, PCR products from the cDNA were also sequenced. First-strand cDNA was synthesized using total RNA and random hexamers with SuperScriptsII reverse transcriptase (Gibco). The cDNA products were amplified with the primers and subcloned PCR products were sequenced as described above. PCR of exon 3 of the *AML1* gene performed using the following flanking intronic , forward/reverse primers; 5'-AGCTGTTTGCAGGGTCCTAA-3'/   5'-GTCCTCCCACCACCCTCT-3'.   The   cDNA products were amplified with the following primers; 5'-GCAGGGTCCTAACTCAATCG-3'/ 5'-GCTCGGAAAAGGACAAGCTC-3', and subcloned PCR products were sequenced as described above.

Two patients had gene rearrangement in exon3 of *AML1*. These samples were used for subsequent DNA sequence analysis and showed C to A, G to A and G to T mutations, respectively. When we further analyzed exon 1 of *AML1*, another three patients who had been exposed to >1.0 Gy were found to have an abnormal band in PCR-SSCP analysis. Exon 1 is located outside the RUNT domain of the *AML1* gene. These might be heteromorphic abnormality because they had the same G to T transition. Two of 6 patients with AB-related AML had abnormality in exon 3 of *AML1*.

The *MLL* gene on the long arm of chromosome 11 was analyzed in 3 patients by the Southern blot method. All of the 6 patients with AB-related AML were shown to have the chromosome breakpoint at 11q22-23 and 11q25, lying outside the *MLL* gene by FISH analysis using several region-specific cosmid probes. Two of the 6 patients had been exposed to >1 Gy. The karyotypes in the 6 patient showed r(11), der(11)(q22), del(11)(q23) and inv(11)(p15q21) in one patient each and add(11)(q21) in 2 patients.

5. Incidence of micronucleus (MN)in lymphocytes and bone marrow cells from healthy AB survivors

It has not yet been clarified whether chromosomal instability is preserved in lymphocytes and bone marrow cells from healthy AB survivors. Chromatid-type aberrations and MN are suitable markers for chromosomal instability because they appear several days after exposure.   Both of bone marrow and lymphocytes samples from these AB-survivors were obtained after 40-50 years after AB radiation exposure. Micronucleus (MN) was evident incidences in the lymphocytes and bone marrow erythroblasts in healthy AB survivors. Furthermore, MN in bone marrow erythrocytes or lymphocytes in 35 and 20 healthy AB survivors without leukemia or hematological disease, respectively, were analyzed using stored smear slides. Cytochalasin B (0.5 μg/ml) was not used for the present lymphocyte MN assay. MN was observed in 1,000 PHA-stimulated lymphocytes, and in 500 bone marrow erythroblasts on smear slides. Cells with MN were scored by microscopy. Exposure doses in most of the healthy AB survivors and patients with AB-related leukemia had been estimated by DS86 dosimetry. Individual approximate exposure doses in the remaining AB survivors were estimated on the basis of the chromosome aberration rate in peripheral blood.

Thirty five AB survivors who had been exposed 2km from the hypocenter in Hiroshima city and whom exposure doses (DS86)had been estimated, were used for MN analysis of bone marrow erythroblasts. Bone marrow smear slide was stained with May-Grünwald Giemsa solution were used for this purpose. Eleven age-matched controls were also studied for comparison. Frequencies of MN in bone marrow erythroblasts were increased by exposure to doses up to 3Gy and after the dose they were slightly decreased. Twenty healthy AB survivors for whom exposure doses had been estimated were used for lymphocyte chromosomal instability. The incidences of MN in 1,000 lymphocytes on peripheral blood smears stained with May-Grünwald Giemsa solution from these AB-survivors were observed in the present analysis. The age-matched non-exposed controls had 1-8 MN per 1,000 lymphocytes, whereas AB-survivors who had been exposed to up to 3 Gy showed a dose-dependent increase of MN in lymphocytes, although small errors in measurement could have occurred in the absence of addition of cytochalacin B. The AB-survivors who had been exposed to more than 4-5 Gy had a slightly lower number of MN than those who had been exposed to 3 Gy. The incidence of MN in lymphocytes clearly increased in a dose-dependent manner. Healthy AB survivors had a higher incidence of MN in lymphocytes and bone marrow erythroblasts, depending on their DS86 exposure dose. These results indicate the presence of AB radiation-induced chromosome instabilities in both bone marrow erythroblasts and lymphocytes of healthy AB survivors.

A healthy AB survivor who developed colon cancer after these examinations was serially analyzed for the chromosome aberration rate and MN incidence. The healthy AB-survivor was a 65 year old woman who had been exposed to 3Gy to Hiroshima bomb, developed colon cancer in 1997. She had been serially observed chromosome aberration rates in lymphocytes for 22 years from 1975 to 1997(Fig.2). Translocation was detected by G-banding using trypsin solution. The frequency of translocation changed from 22-35% during this period, and then increased rapidly to 46 % at the time of colon cancer development. The incidences of both spontaneous chromosome aberrations such as chromatid breaks and hyperploidy and MN were higher than in the control. Chromatid-type aberrations such as breaks and gaps were also serially followed up during this period. Spontaneous chromosome aberrations, including the total rates of chromatid-type aberrations and hyperdipolidy and hyperploidy, are shown in Fig.2. The frequency of spontaneous chromosome aberrations remained at around 6-14% in 1975-1991 and it increased at the time of cancer development. MN incidence in lymphocytes was also followed up serially during this period and showed a similar pattern (Fig.2). The incidence of MN per 1,000 lymphocytes was 4-7 and increased to 16 at the time of cancer development. Serial observations in an AB-survivors showed that all of the cytogenetic markers of chromosomal instability, including translocations, spontaneous chromosome aberrations and MN increased at the time of colon cancer development.

6. Plausible mechanisms for development of chromosome instability in AB-related leukemia and healthy AB survivors

In vitro culture studies have demonstrated radiation-induced chromosome instability in human and rodent cells after several cell divisions (Kadhim et al., 1992; Holmberg et al. 1993). These chromosome abnormalities are known as delayed chromosome aberrations (Tanaka and Ihda 2008; Tanaka et al. 2008). MN and chromatid-type aberrations are suitable cytogenetic markers for detecting chromosome instability because they are eliminated by cell division and appear only for a short time after radiation exposure, being associated

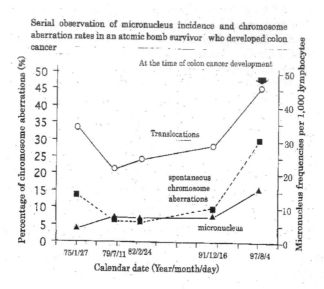

Fig. 2. Serial observation of MN incidence and chromosome aberration rates in an AB survivor who developed colon cancer

with the approximately 3-year life span of lymphocytes. Chromosome aberrations appear a long time after radiation exposure,and are considered to be an indirect radiation exposure. The frequency of translocations found in AB-survivors depends on the radiation exposure dose, but such aberrations are not suitable for proving chromosomal instability. AB-survivors who were exposed to higher radiation doses had a higher number of MN in both erythroblasts and lymphocytes 30-40 years after the bombing. The incidence of MN in lymphocytes clearly increased in a dose- dependent manner. However, conflicting result has been reported for dicentric chromosomes in peripheral lymphocytes in healthy AB survivors (Hamasaki *et al.* 2009). These results provide clues as to why the individuals who were exposed to radiation showed higher chromosomal instability.

First, it can be hypothesized that a small percentage of DNA damages is not repaired completely and persists for a long time after exposure to radiation. At several days after exposure, un-repaired sites are recognized by DNA-binding proteins in an error-prone manner. The key issue in this model is the type of mechanism responsible for long-term preservation of cells with un-repaired DNAs (Suzuki *et al.*, 2003). Gene recombination associated with chromosomal instability will occur at the time. A second hypothesis is that inflammation occurs in the organs of AB-survivors exposed to higher radiation doses(Neriishi *et al.*, 2003). Macrophages, NK-cells, lymphocytes, monocytes release active oxygen species into areas of inflammation. Such oxygen stress can induce chromatid-type of aberrations and MN in lymphocytes located in the vicinity. Similar findings were observed in mice experiments (Lorimore *et al.*, 2001). Interestingly, the most common abnormality for delayed chromosome instability observed in mice were chromatid-type aberrations (Watson *et al.* 1996, 2001; Ullrich and Davis 1999), which imply released reactive oxygen species or secreted factors related to the chromosome instability. The present serial analyses of MN and spontaneous chromosome aberrations in a patient with AB-related colon cancer also revealed that chromosome instabilities increased rapidly at the time of cancer development. In this case also, it is considered that phagocytic cells that had accumulated in cancer-prone tissues released active oxygen species, which induced chromosome aberrations and MN in lymphocytes. Of course, however, the first hypothesis also still remains to be examined.

AB-related leukemias showed more complex chromosome aberrations than *de novo* leukemias. Twenty-seven of the 33 patients with AB-relatedAML showed complex chromosome aberrations involving more than three chromosomes including chromosomes 11 and 20, and the number of chromosomes per cell in AB-relatedleukemias was 3.69, compared with 0.93 of *de novo* leukemia. Only one of the 33 patients (3.1%) had a normal karyotype, compared with 44.1% of *de novo* leukemia. It is considered that phagocytic cells that had accumulated in cancer-prone tissues released active oxygen species, in turn inducing chromosome aberrations in bone marrow leukemic cells. Of course, the hypothesis that malignant clones might have developed from hematopoietic stem cell with higher chromosomal instability remains to be investigated.

7. Different mechanisms for development of AB-related AML and *de novo* AML

Most of the 33 patients with AB-related AML had a preceding long term MDS stage. Therefore, no leukemia type-specific translocations of such as t(8;21), t(15;17), t(11;A)(q23;A) and inv(16) were found in these 33 patients. The best approach is to compare types of chromosome aberrations between the patients with AB-related MDS and those with AB-related AML developed from MDS. Translocations on chromosome 11 at 11q13 to 11q23, and deletion/loss of chromosome 20 were frequently found in AB-related leukemias (Table 1). The fact that AB-related leukemias did not have chromosomal breakpoints in the *MLL* gene indicates that AB-related leukemia develops from MDS. MDS has shown to have a chromosome breakpoint at 11q22, not 11q23 (Tanaka *et al.*, 2001). Most cases of 11q23 AML do not have a MDS stage (Tanaka *et al.* 2001). The reason why AB-related leukemia did not have a translocation breakpoint within the *AML1* gene on 21q22 would be the same as that for11q22 abnormality in MDS. It is well known that deregulation of the *AML1* gene results from gene mutations as well as chromosomal translocations such as t(8;21), t(3;21), t(12;21) and so on (Harada *et al.* 2003).

It has been reported that secondary MDS developing after chemotherapy has been shown to have a high incidence of *AML1* gene mutation and that MDS shows a higher incidence of mutations in the *AML1* gene (Harada *et al.*, 2003),in which the frequencies of AML1 point mutations were 17% (15/88 cases) in sporadic MDS/AML and 50% (11/22 cases) in secondary MDS/AML (4/20). In present study, 2 of 6 patients with AB-related AML had point mutations in exon 3 on the *AML1* gene. On the other hand, PCR-SSCP analysis of non-AB-related AML patients has shown that only 2 showed transformation from MDS and that none of 14 MDS patients diagnosed as having RAEB-T had abnormalities(i.e. 2 of 30 non-AB-related AML patients showed transformation)(Harada and Harada, 2009), thus indicating that AB-related

AML might have a higher incidence of *AML1* gene mutation. AMLs that developed in residents at living near the Nevada nuclear weapons testing site were reported to have chromosome aberrations involving the *AML1* gene at 21q22 on chromosome 21 (Hromas *et al.*, 2000). The two of the present 6 patients with AB-related AML who had been exposed to more than 1Gy had point mutations in exon 3, which encodes the RUNT domain of AML1(CBFβ 2) protein. It will be necessary to compare the break points in the *AML1* gene (exons 1 and 3) and the base changes in the DNA sequences between these two AB-related leukemias and the published DNA sequences in MDS. AML 1(CBFβ 2) protein having point mutation could not bind with both histon deacetylation (HDAC) and histon acetylation HAT, which results in suppression of transcription and transformation of MDS to AML (Ding *et al.*, 2009). C- terminal *AML1* point mutations (exons 3-5) were exclusively found in sporadic MDS/AML, while N-terminal(exons 6-8) mutations were found in chemotherapy related-secondary MDS/AML (Harada *et al.*, 2003). Therefore the role of exon 3 mutations of *AML1* gene in the pathogenesis of radiation-induced leukemia has been questioned.

These results suggest that AB-related leukemia is derived from abnormal pluripotent hematopoietic stem cells, which are preserved for a long term and show higher genetic instability such as microsatellite instability (Nakanishi *et al.*, 2001)(Fig.3), whereas *de novo* AML develops from committed hematopoietic stem cells and shows simple chromosome aberrations. t(8;21)-positive leukemic cells arise from committed stem cells, on the other hand del(20), t(11,?) might be derived from pluripotent stem cells, as in the case for deletion of chromosome 5/monosomy 5 at the pluripotent stem cell level (Nilson *et al.*, 2007). Both of these leukemias developed from MDS. Thus the *AML1* gene might be activated only at the hematopoietic stem cell level.

Genetic instability is preserved in hematopoietic stem cells damaged by radiation.

As most AB-leukemias developed from MDS,it is more appropriate to compare AB-leukemia with *de novo* AML having a MDS stage before development. Observation of AB-leukemias in bone marrow smears showed that their hematological features were similar to those of MDS found in aged patients. Recent leukemia study showed that a higher incidence of MDS was observed in Nagasaki AB survivors (Iwanaga *et al.*, 2011), which implicates that AB-related MDS would have more chance to acquire a higher chromosome instability in long-term after exposure to AB-radiation. This also suggested that AB-radiation might induce earlier development of MDS. Chromosome aberrations found in AB-related leukemia included a higher frequency of del(20) and 11q22-11q23 abnormalities than those in *de novo* leukemias. Monosomy and deletion of chromosome 5 (-5/del(5)) occurred at almost the same frequencyas those in *de novo* leukemias. On the other hand monosomy 7 and deletion of chromosome 7(-7/del(7)) occurred at lower frequencies in AB-related leukemia than in *de novo* leukemias (Arif *et al.*, 1997). This indicated that the type of chromosome aberrations in AB-related leukemias showed different from those in *de novo* AML and MDS. AB radiation might induce different type of cancer as well as cancer development with a shorter latency. A similar finding has been observed using analyses of array CGH and cell surface marker by FACS sorting for low-dose-rate radiation-induced murine leukemias, lymphomas and tumors by The Institute for Environmental Science (Tanaka *et al.*, 2007; Hirouchi *et al.*, 2009, 2011).

In conclusion, chromosome and oncogene features revealed in present study might be related to the higher incidence of AML transformed from MDS in AB survivors a long time

Cytogenetic Instabilities in Atomic Bomb-Related Acute Myelocytic Leukemia Cells and in Hematopoietic
Cells from Healthy Atomic Bomb Survivors

121

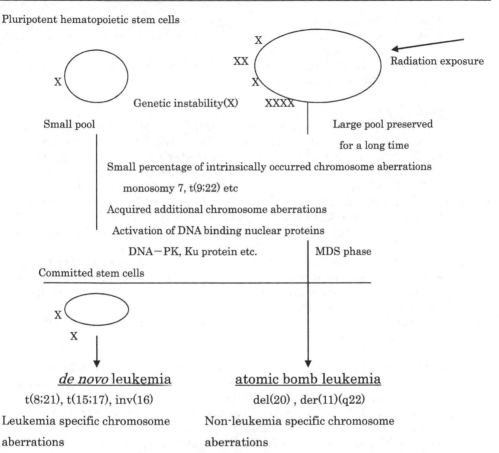

Fig. 3. Hypothetical schema of the development of radiation-induced leukemia from
pluripotent hematopoietic stem cells

after exposure, and these findings are important for understanding the mechanisms
responsible for radiation-induced leukemia.

8. Acknowledgements

I would like to thank Dr. Hideo Tanaka, Department of Hemato-oncology, Asa-kita
Hospital, Hiroshima, for *AML1* gene mutation analysis, Prof. Nanao Kamada, Hiroshima
Atomic Bomb Relief Foundation, Hiroshima, for general comments and discussion and Drs.
Taichi Kyo and Hiroo Dohy, Hiroshima Red Cross Hospital for clinical information.

9. References

Arif M., Tanaka K., Chendil D., Asou H., Kyo T., Dohy H., Kamada N. (1996) Hidden
monosomy 7 in acute myeloid leukemia and myelodysplastic syndrome detected
by interphase fluorescence in situ hybridization. *Leukemia. Res.* Vol. 20, pp.709-716.

Ding Y., Harada Y., Imanaga J., IMURA A., HARADA H.(2009) AML1/RUNX1 point mutation possibly promotes leukemic transformation in myeloproliferative neoplasms. *Blood*Vol. 114, No.25, pp.5201-5205.

Hamasaki K., Kusunoki Y., Nakashima E., Takahashi N., Nakachi K., Nakamura N., Kodama Y. (2009) Clonality expanded T lymphocytes from atomic bomb survivors in vitro show no evidence of cytogenetic instability. *Radiat. Res.* Vol. 172, No 2, pp.234-243.

Harada H., Harada Y., Tanaka H., Kimura A., Inaba T. (2003) Implication of somatic mutations in the *AML1* gene in radiation-associated and therapy-related myelodysplastic syndorome / acute myeloid leukemia. *Blood* Vol. 101, pp. 673-680.

Harada Y. and Harada H.(2009) Molecular pathways mediating MDS/AML with focus on AML1/RUNNX1 point mutations. *J Cell Physiol.* Vol. 220, No.1, pp.16-20.

Hirouchi T., Takabatake T., Yoshida K., Nitta Y., Nakamura M., Tanaka S., Ichinohe K., Oghiso Y., Tanaka K., (2009) Upregulation of c-myc gene accompanied by PU.1 deficiency in radiation-induced acute myeloid leukemia in mice. *Exp. Hematol.* Vol. 36, pp. 871-881.

Hirouchi T., Akabane M., Tanaka S., Tanaka I. B., Todate A., Ichinohe K., Oghiso Y., Tanaka K. (2011) Cell surface marker phenotype and gene expression profiles of murine radiation-induced acute myeloid leukemia stem cells are similar to those of common myeloid progenitors. *Radiat. Res.* Vol.176, pp. 311-322.

Holmberg K., Falt S., Johansson A., Lambert A. (1993) Clonal chromosome aberrations and genomic instability in X-irradiated human T-lymphocyte cultures. *Mutat. Res.* Vol. 286, pp. 321-330.

Hromas R., Shopnick R., George H., Bowers C.,Varella-Garcia M., Richkind K.(2000) A novel syndrome of radiation-associated acute myeloid leukemia involving AML1 gene translocation. *Blood* Vol. 15,pp. 4011-4013.

Iwanaga M, Hsu W-I., Soda M., Takasaki Y., Tawara M., Joh T., Amenomori T., Yamamura M., Yoshida Y., Koba T., Miyazaki Y., Matsuo T., Preston D. L., Suyama A., Kodama K., Tomonaga M. (2011) Risk of myeloid dysplastic syndrome in people exposed to ionizing radiation: a retrospective cohort study of Nagasaki atomic bomb survivors. *J. Clin. Oncol.* Vol. 29, No.4, pp. 428-434.

ISCN (2009) An International System for Human Cytogenetic Nomenclature. Eds. Shaffer L. G., Slovak M. L. and Campbell L. J., Karger, Basel

Kamada N. and Tanaka K. (1983) Cytogenetic studies of hematological disorders in atomic bomb survivors . *Radiation-induced Chromosome Damage in Man,* (T. Ishihara and M. S. Sasaki Eds.), pp.455–474, Alan R. Liss, New York

Kamada N. and Tanaka K, Oguma N., Mubuchi K.(1991) Cytogenetic and molecular changes in leukemia among atomic bomb survivors. *J. Radiat Res.* Vol.32,Suppl.2, pp.257-265.

Lorimore S. A., Coates P. J., Scobie G. E., Milne G., Wright E. G. (2001) Inflammatory-type responses after exposure to ionizing radiation *in vivo*; a mechanism for radiation-induced bystander effects? *Oncogene* Vol.20, pp. 7085-7095.

Nakanishi M.,Tanaka K., Shintani T., Takahashi T., Kamada N.(1999) Chromosomal instability in acute myelocytic leukemia and myelodysplastic syndrome patients among atomic bomb survivors. *J. Radiat. Res.* Vol.40,pp.159-167.

Cytogenetic Instabilities in Atomic Bomb-Related Acute Myelocytic Leukemia Cells and in Hematopoietic
Cells from Healthy Atomic Bomb Survivors

123

Nakanishi M., Tanaka K., Takahashi T., Kyo T., Dohy H., Fujiwara M., Kamada N.(2001) Microsatellite instability in acute myelocytic leukemia derived from A-bomb survivors. *Int. J. Rad. Biol.* Vol. 77, pp. 687-694.

Neriishi K, Nakashima K, Delongchamp R. R. (2003) Persistent subclinical inflammation among A-bomb survivors. *Int. J. Radiat. Res.* Vol. 77, pp.475-482.

Nilson L., Edén P., Olsson E., Månsson R., Åstrand-Grundström I., Storömbeck B., Theilgand-Mönch K., Andeson K., Hast R., Hellstörme-Lindberg E., Samuelsson J., Berg G., Nerlov C., Johansson B., Sigvardsson M., Borg Å, Jacobsen S. E. W.(2007) The molecular signature of MDS stem cells supports a stem-cell origin of 5q-myelodysplastic syndromes. *Blood* Vol. 110, No.8, pp.3005-3014.

Pierce D. A., Shimizu Y., Preston D. L., Vaeth M., Mabuchi K. (1996) Studies of the mortality of atomic bomb survivors. Report 12, Part 1. Cancer: 1950-1990. *Radiat. Res.* Vol.146, No.1, pp.1-27.

PrestonD. L., Kusumi S., Tomonaga M., Izumi S., Ron S., Kuramoto A., Kamada N., Dohy H., Matsuo T., Masui T.(1994) Cancer incidence in atomic bomb survivors. Part 1. Leukemia, lymphoma and multiple myeloma, 1950-1987. *Radiat. Res.* Vol. 137(Suppl.) S86-97.

Richardson D., Sugiyama H., Nishi N., Sakata R., Shimizu Y., Gran E. J., Soda M., Hsu W. L., Suyama A., Kodama K., Kasagi F. (2009) Ionizing radiation and leukemia mortality among Japanese Atomic Bomb Survivors, 1950-2000. *Radiat. Res.* Vol.172, No.3, pp. 368-382.

Suzuki K., Ojima M., Kodama K., Suzuki M., Oshimura M., Watanabe M. (2003) Radiation-induced DNA damage and delayed induced genomic instability. *Oncogene* Vol.22, pp. 6988-6993.

Tanaka I. B., Tanaka S., Ichinohe K., Matsushita S., Matsumoto T., Otsu Y., Sato F. (2007) Cause of death and neoplasia in mice continuously exposed to very low-dose rates of gamma rays. *Radiat. Res.* Vol. 167, pp. 417-437.

Tanaka K., Hoshi M., Kamada N., Sawada S. (1994) Effects of 252 Cf neutrons, transmitted through an iron block on human lymphocyte chromosome. *Int. J. Radiat. Biol.* Vol. 66, pp.391-397.

Tanaka K., Arif M., Eguchi M., Kyo T., Dohy H., Kamada N.(1997) Frequent jumping translocations of chromosomal segment involving the ABL oncogene alone or in combination with CD3-MLL genes in secondary leukemias. *Blood*Vol. 89, pp. 596-600.

Tanaka K., Gajendiran N., Endo S., Hoshi M., Kamada N. (1999) Neutron energy-dependent initial DNA damage and chromosomal exchange. *J. Radiat. Res.* Vol.40, Supl. pp. 36-44.

Tanaka K. , Eguchi M., Eguchi-Ishimae M., Hasegawa A., Ohgami A., Kikuchi M., Kyo T., Asaoku H., Dohy H. and Kamada N. (2001) Restricted chromosome breakpoint sites on 11q22-23.1 and 11q25 in various hematological malignancies without *MLL/ALL-1* gene rearrangement. *Cancer Genet. Cytogenet.* Vol. 124, pp. 27-35.

Tanaka K. and Ihda S. (2008) Radiation-induced chromosome instability and luekemogenesis in human. *J. Genetic Toxicol.* Vol. 1, No. 1,pp.1-19.

Tanaka K., Kumaravell T. S., Ihda S., Kamada N. (2008) Characterization of late –arising chromosome aberrations in human B-cell lines established from alpha-ray- or gamma-ray-irradiated lymphocytes. *Cancer Genet. Cytogenet.* Vol 187, pp. 112-124.

Ullrich R. L. and Davis C. M. (1999) Radiation-induced cytogenetic instability *in vivo*. *Radiat. Res.* Vol. 152, pp. 170-173.

Watson G. E., Lorimore S. A., Wright F. G. (1996) Long-term *in vivo* transmission of alpha-particle-induced chromosomal instability in murine haemopoietic cells. *Int. J. Radiat. Biol.* Vol. 69, pp. 175-182.

Watson G. E., Pocock D. A., Papworth D., Lorimore S. A., Wright E. G. (2001) *In vivo* chromosomal instability and transmissible aberrations in the progeny of hematopoietic stem cells induced by high- and low-LET radiations. *Int. J. Radiat. Biol.* Vol. 77, pp. 409-417.

8

Chromosomes as Tools for Discovering Biodiversity – The Case of Erythrinidae Fish Family

Marcelo de Bello Cioffi[1], Wagner Franco Molina[2],
Roberto Ferreira Artoni[3] and Luiz Antonio Carlos Bertollo[1]
[1]*Universidade Federal de São Carlos*
[2]*Universidade Federal do Rio Grande do Norte*
[3]*Universidade Estadual de Ponta Grossa*
Brazil

1. Introduction

Biodiversity or biological diversity is the diversity of life, extant or extinct. All of the biodiversity found on Earth today consists of many millions of distinct biological species and is the product of more than 4 billion years of evolution. Although the origin of life has not been correctly determined by science, some evidence suggests that life may already have been well-established only a few hundred million years after the formation of the Earth. Estimates of the number of extant global macroscopic species vary from 2 million to 100 million, with a best estimate of approximately 13–14 million, and the vast majority is represented by insects. However, biodiversity is not evenly distributed; rather, it varies greatly across the globe as well as within regions. A "biodiversity hotspot" can be defined as a region with a high level of endemic species, and while they can be found all over the world, the majority of them are forest areas, and most are located in the tropics (Myers, 1988).

In fact, it is not a coincidence that the world's biodiversity hotspots are also the centers of evolutionary change for numerous species. Evolution produces biodiversity, and in turn, a more diverse biological environment creates more selective pressures, which drive evolution. The biodiversity of a specific region is often measured by determining the number of species found there. Although biodiversity concept with its complex mutual evolutionary interrelationships is not merely a species inventory, complete list of reliably identified species is a basic prerequiste for addressing various biodiversity issues. Therefore, for an accurate assessment of biodiversity, it is first necessary that a correct definition of the term species as well as the methods to differentiate between species be considered. However, to identify species requires i) appropriate species concept, ii) informative diagnostic characters that correctly separate species, i.e diagnosable unit. The problem of species concepts in ichthyology has been extensively dicussed (Nelson 1999, Ruffing et al., 2002, Mooi & Gil, 2010) and compete for acceptance. One of them, and also the one used in this book chapter, is the "evolutionary species concept" (Mayr 1942), which defines a species not only according to appearance but as members of populations that actually or potentially interbreed in nature in such a way that they are necessarily reproductively isolated from others, has its own independent evolutionary fate and its own historical tendencies, thus representing separate evolutionary lineage.

However, species can be identified by means of diagnostic characters at different levels of organism/genome organization (Ráb et al., 2007). One of these levels, genetic diversity, is composed of the diversity in organisms and populations arising from genetic and genomic variants. Genetic variation is a fundamental characteristic of most biota. Recently, advances in genetic technologies have permitted deep investigations of the genetic variation among distinct living or even extinct groups. Such methodologies have enabled the exploration of various questions related to biodiversity and its conservation (i.e., determination of genetic variability, population delineation, characterization of the germplasm, genotoxicity, comparative genomic studies, forensic analysis, phylogeographic patterns and evolutionary inferences). Among these approaches, cytogenetic studies have proved to be a useful tool in several cases by identifying the chromosomal characteristics of the genomes of a species. Indeed, in some situations, changes in chromosome number and structure have been correlated with a number of novel morphological and environmental traits leading to habitat divergence and adaptation (Hoffmann & Rieseberg, 2008). In this chapter, we advocate the view that, as compared to various other genetic markes, cytogenetics can reveal set of characters, many times diganostic ones, that are not accessible by other methods and thus explore another level of genetic component of biodiversity. We exemplify this view using the Erythrinidae fish family in which classic and molecular cytogenetic techniques were useful to compare the degree of chromosomal diversity over a species geographical range, providing important tools for evolutionary and taxonomic studies, besides improving the knowledge of the genome diversification and the biological biodiversity.

2. Cytogenetics and biological investigation

For many years, classical conventional karyotyping methods have been used to determine chromosome number and morphology as well as the presence of morphologically differentiated sex chromosome systems in many animal and plant species. This approach has contributed significantly to the present knowledge of chromosomal diversity and/or stability among many distinct taxa. The advent of chromosome banding techniques (i.e., C-, G-, R-, Q- and H-banding and AgNORs, DAPI and CMA_3 staining) allowed for the differentiation of specific regions along chromosomes, and many important karyotypic changes could be demonstrated with such methodologies. The recent combined use of cytogenetic, genetic and refined molecular studies has helped us to understand the connection between genomic organization and the functional biology of chromosomes, species adaptation and species survival, and this has begun a new era of evolutionary genomics and phylogenomics.

Major advances in cytogenetics arose in the last two decades with the application of *in situ* detection of DNA sequences on chromosomes, parts of chromosomes or even whole genomic DNA. The earliest *in situ* hybridizations, performed in the late 1960s, were not fluorescent but instead utilized probes labeled with radioisotopes (i.e., ^{3}H, ^{35}S, ^{125}I and ^{32}P). Since the beginning of the 1980s, probes started being labeled with non-radioactive molecules. Although several methods based on enzymatic reactions using alkaline phosphatase, beta-galactosidase or horseradish peroxidase were available, the most commonly used method in the subsequent years was based on the utilization of fluorescent elements; therefore, the technique was named (Fluorescence *In Situ* Hybridization) (FISH) (Pinkel et al. 1986). In the FISH technique, probe detection and experimental results are based on the observation of fluorescent signals through an epifluorescence microscope. FISH emerged as a useful alternative to older hybridization methods because the fluorescent systems had more precise definition of the hybridization signals as compared to the radioactive or enzymatic methods (Gall & Pardue, 1969). Indeed, FISH technology, where a DNA probe is hybridized to its complementary sequences on previously fixed and denaturated chromosomal DNA preparations, appeared to be superior to previous *in situ* technologies, as it allowed the simultaneous use of different fluorescence systems for multi-probe analysis and the detection of more than one target sequence simultaneously. The analysis of FISH results requires the use of an epifluorescence microscope coupled with digital cameras connected directly to a computer for the image acquisition. Additionally, nowadays some computer softwares can also be used for image manipulation, increasing the final quality of the figures.

One of the most important applications of the FISH technique has been its use in the physical mapping of DNA sequences on chromosomes. Its development has led to the advancement of chromosome studies not only for physical mapping and genome analyses but also as a tool for evolutionary and phylogenetic studies. With FISH, it is possible to map the location of DNA sequences across related species and genera to show not only their probable conservation but also their diversification throughout evolutionary processes. Thus, the advent of FISH allowed the transition from the "classical" (black-and-white) to the "molecular cytogenetic" (color) and combined with genomic data recently to phylogenomic era, which allowed the integration of molecular information of DNA sequences with its physical location along chromosomes and in genomes (Schwarzacher, 2003) **(Figure 1).**

The simple characterization of the karyotype in some species may be sufficient to identify intra- and inter-cytotype variants and to characterize species. However, in most cases, just the the karyotype descriptions appear to be inconclusive when not coupled with other methods capable of generating more accurate chromosomal markers for cytotaxonomy or phylogenetic applications. In part, this stems from the inability to discern either the mechanisms involved in karyotype evolution of some groups (homologies *versus* convergences) or the genesis of novel chromosomal structures. In some groups, such as Perciformes fishes, karyotypes and cytogenetic aspects associated with the chromosomal structure, identified by conventional cytogenetic techniques, show a vast number of species sharing the same karyotypic patterning, which restricts their use for taxonomic and phylogenetic inferences (Molina, 2007; Motta-Neto et al., 2011).

In contrast, the molecular organization and cytogenetic mapping of many genes might be a significant data set for the characterization of particular segments of biota, providing very important information for phylogenomics. Remarkably, a substantial fraction of any eukaryotic genome consists of repetitive DNA sequences including satellites, minisatellites, microsatellites and transposable elements (Jurka et al., 2005). These repetitive DNA sequences are thought to arise through many mechanisms, from direct sequence amplification by unequal recombination of homologous DNA regions to the reverse flow of

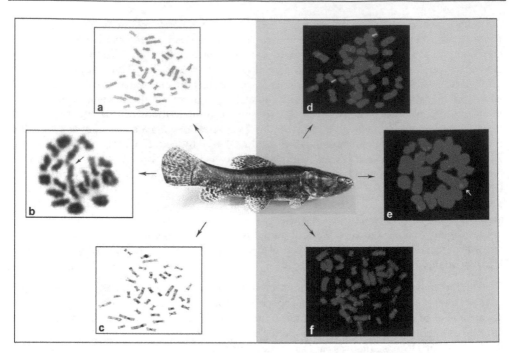

Fig. 1. Conventional (a, b and c) x molecular (d, e, and f) cytogenetic analyses in chromosomes of males of the karyomorph D of the fish *Hoplias malabaricus* (2n = 39, X_1X_2Y sex chromosomes) highlighting the transition from the "black-and-white" to the "color" era. Giemsa-stained mitotic chromosomes (a) and diakinesis/metaphase I meiotic cell (b); C-banded (c); 5S rDNA (green) and satellite 5S*Hind*III-DNA (red) hybridized to mitotic chromosomes (d); diakinesis/metaphase I meiotic cell displaying 18S rDNA sites (red) in the synapsed chromosomes (e); simple sequence repeat (GA)$_{15}$ (red) hybridized to mitotic chromosomes. The arrows indicate the sex trivalent.

genetic information using an intermediate RNA molecule. Due to the hypervariability of tandem repeats, such genomic segments are highly polymorphic and considered to be good molecular markers for genotyping individuals and populations (Jeffreys et al., 1985). The repetitive fraction of the genome was long considered to be "junk DNA" with no clear function, which was reinforced by indications that these sequences were not transcribed in eukaryotes (Doolittle & Sapienza, 1980). However, accumulated data from eukaryotic species of diverse taxonomic origins have challenged this view over the past few years (Bonaccorsi & Lohe, 1991), supporting a major role of repetitive DNA sequences in the structural and functional evolution of genes and genomes in a variety of organisms (Biémont & Vieira, 2006).

Today, many studies have been conducted using dispersed or *in tandem* repetitive DNA sequences as probes for FISH cytogenetic mapping in distinct living organisms; these sequences include simple sequence repeats **(Figure 2a)**, satellite DNA **(Figure 2b)**, BAC clones **(Figure 2c, e)**, and rRNA genes **(Figure 2d, f)**. In general, these probes provide highly visible signals due to their abundant repetition and distribution in the genome, and they

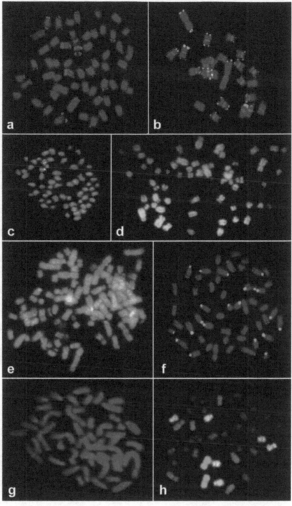

Fig. 2. Fluorescence *in situ* hybridization to metaphase chromosomes of distinct organisms using different probes. (a) simple sequence repeat (CA)$_{15}$ (red) in the fish *Leporinus elongatus*; (b) Centromeric STAR-C tandem repeat (red) and subtelomeric X43.1 tandem repeat (green) in the plant *Silene latifolia*; (c) BAC-FISH in the moth *Biston betularia* using BAC clones of sulfamidase (green) and lrtp (red); (d) 45 rDNA (red) in the eagle *Spizaetus tyrannus*; (e) BAC-FISH in the hawk *Leucopternis albicollis* using BAC clones derived from a *Gallus gallus* microchromosome; (f) 5S rDNA (green) and 18S rDNA (red) in the fish *Erythrinus erythrinus;* (g) chromosome paint probe of human chromosome no. 7 (red) hybridized to the chromosomes of the monkey *Alouatta fusca*; (h) multi-color FISH hybridization of cat painting probes hybridized to the chromosomes of the leopard *Panthera pardus.* The FISH image (b) is courtesy of Eduard Kejnovsky (Academic Science of Czech Republic, Czech Republic), (c) of František Marec (Biology Centre of the Academy of Sciences of the Czech Republic, Czech Republic), (d and e) of Edivaldo HC de Oliveira (Instituto Evandro Chagas, Brazil), and (h) of Vladimir Trifonov (Institute of Chemical Biology and Fundamental Medicine, Russia).

might even generate a unique FISH karyotype for each species (e.g., Badaeva et al., 2007), enabling an evolutionary and phylogenetic view of related species. In contrast to functional genes, repetitive DNA sequences are thought to have evolved under different conditions, escaping from the selective pressures that act on the non-repetitive segments (Charlesworth et al., 1994). In this sense, they represent good chromosomal markers to detect recent differentiation events.

In addition, Genome *In Situ* Hybridization (GISH) and chromosome painting are new and useful tools for investigating biodiversity. GISH allows for the comparison of genomes using the genomic DNA of one organism as a probe for the DNA of another organism. This method offers new perspectives in phylogenetic and systematic studies by determining and testing hypotheses of genomic relatedness between species. In turn, chromosome painting has also provided significant support to comparative cytogenetics by highlighting chromosomal changes that took place during the evolution of a species. DNA probes covering an entire chromosome can be developed using chromosome sorting with microdissection-based or flow-sorting methods. Such whole chromosome probes (wcp) allow the tracking of homologous and/or segments of chromosomes among related species and have been a powerful tool for evolutionary studies being used to identify homologous chromosome segments among different species, rearrangements and thereby karyotype differentiation. In recent years, complete karyotypes of many animal and plant species have been analyzed by chromosome painting, which have added to our understanding of genomic reorganization and chromosome evolution (Griffin et al., 2007, Ferguson-Smith & Trifonov 2007; Teruel et al., 2009; Yang & Graphodatsky, 2009; Cioffi et al.,2011a; Pokorná et al., 2011) **(Figure 2g, h)**.

In summary, the development and improvement of cytogenetic FISH analyses have substantially expanded the methods of chromosome studies and have played an important role in the precise characterization of the structure of genomes. The current availability of an ever increasing number of completely sequenced eukaryotic genomes has opened new "avenues" for advancing cytogenetics. Coupled with the application of bioinformatics, the integration of chromosome analysis and genomic data represents promising tools for the future of cytogenetics. However, classical information regarding chromosome number and morphology and banding data is not outdated and therefore should still be useful to elucidate a range of both basic and applied aspects, ranging from cytotaxonomy to karyotype evolution.

In fact, a number of groups of organisms show a high diversity of species. However, this diversity is frequently distributed in a phylogenetically uneven way. Among fish taxa, for example, an order might contain anywhere from two to over ten thousand species (Nelson, 2006). In groups widely distributed and difficult to access, diversity is only estimated, and the number of species identified each year continues to dramatically increase. Many of these species are known to be endemic; thus, they have a crucial, pivotal role in biological conservation. Indeed, issues related to cryptic biodiversity and its correct identification continue to demand constant attention.

3. Neotropical fish and biodiversity

3.1 Fish: Diversity and functional role in evolutionary studies

Fish exhibit the greatest biodiversity among the vertebrates, making this group extremely attractive to study a number of evolutionary questions. The term "fish" most precisely describes any non-tetrapodal craniate that has gills throughout life and whose limbs, if any, are in the shape of fins. However, fish do not constitute a monophyletic group but are instead a paraphyletic collection of taxa including hagfish, lampreys, sharks, rays and the finned bony fish. The latter is by far the most diverse group and is well represented in freshwaters, while the others are predominantly marine groups. Nelson (2006) suggested the presence of almost 34,500 fish species out of the almost 55,000 recognized living vertebrate species.

Generally, each continent has a distinctive freshwater fish fauna, and the observed patterns of fish distribution are the result of physical barriers disrupting past fish dispersal and different temperature adaptations amongst the various groups. Most species occur in the tropical and subtropical regions, and there is an overall reduction in diversity in temperate and polar regions (Lévêque et al., 2007). Specific aspects of the spatial distribution of this group, such as subdivisions according to biogeographical barriers and biological aspects such as length of life, population size, degree of mobility, behavior patterns, and aspects of sex determination are reflected in their chromosomal patterning. The wide spectrum of mechanisms for reproduction, sex determination and sexual differentiation in fish species also illustrates the plasticity of their genomes, with many species exhibiting hermaphroditism and some even changing sex at a specific stage in their life cycle. Indeed, fish show a range of sex determination mechanisms, from male or female heterogametic sex determination to environmental sex determination (Devlin & Nakayama, 2002). It has been suggested that all this diversity might be related to the fact that fish genomes seem to undergo genetic changes more rapidly than in other vertebrate groups (Venkatesh, 2003).

Although fish have traditionally been the subject of comparative evolutionary studies, they have now drawn attention as models in genomics and molecular genetics research, and there are many ongoing or completed genome sequencing projects, including those for the catfish *Ictalurus punctatus,* the rainbow trout *Oncorhynchus mykiss,* the Atlantic salmon *Salmo salar,* the three-spined stickleback *Gasterosteus aculeatus,* the Nile tilapia *Oreochromis niloticus,* the two pufferfish *Takifugu rubripes* and *Tetraodon nigroviridis,* the platyfish *Xiphophorus maculatus,* the medaka *Oryzias latipes,* the spined loach *Cobitis taenia* and the popular zebrafish (*Danio rerio*), which is a commonly used model organism for studies of vertebrate development and gene function (Mayden et al., 2007).

3.2 The Neotropical region and freshwater fish biodiversity

Concerning all of the biodiversity on Earth, the Neotropical region has the largest repository of genetic information, and its biodiversity has an enormous economic importance in addition to its ecological relevance. The number of freshwater fish species in the world is estimated to be approximately 15,000. Although a substantial component of the Neotropical fish fauna is still unknown, approximately 6,000 freshwater fish species are found in this region (Reis et al., 2003), which corresponds to approximately 45% of all freshwater fish species in the world (Oliveira et al., 2007).

The hydrographic system that drains the Neotropical region is highly branched covering a large and ecologically diverse area and containing an extremely diverse fish fauna, one of the world's richest in number of species. Phylogenetic and biogeographic patterns indicate that, in most groups of Neotropical fishes, diversification occurred incrementally over large spatial and temporal scales, with speciation occurring over much of the continental platform and requiring tens of millions of years. Complementary, vicariance and species dispersal processes have profoundly influenced the formation of new species and the taxonomic composition of regional biotas. Together, these processes interact in a complex duet of Earth history events and biological diversification (Albert & Reis, 2011). Therefore, the exploration of Neotropical fish evolution requires a multidisciplinary approach to gain a more complete understanding, and cytogenetics has contributed to it as an important tool to support/correct systematic studies and the corresponding taxonomical constructions.

3.2.1 Cytogenetic studies in Neotropical fishes

Among Neotropical fish, there are many nominal species (i.e. group of individuals appointed as taxonomically unique and but not necessarily validated by additional biological and genetic studies) with a large geographic distribution that are found in different river basins isolated by millions of years. In this region, small and widely distributed fish that inhabit small streams with limited opportunity to migrate tend to possess an increased rate of speciation and form a "species complex".

In such ecological systems, cytogenetic studies have made important contributions toward a better understanding of Neotropical fish fauna, showing that many local populations have different chromosomal characteristics. However, most of these studies utilized classical techniques involving conventional staining and simple banding procedures such as C-banding, the detection of nucleolar organizer regions (AgNOR) and fluorochrome staining techniques. Despite this apparent limitation, the results have provided important data that have revealed interesting and significant components of this biodiversity, which have helped to elucidate the evolutionary pathways of distinct fish groups. A number of cases of species complexes, populational polytypy, the presence of B-chromosomes, diverse sex heteromorphic chromosome systems and spontaneous polyploidization have been reported in various Neotropical fish species.

Karyotype data have shown an extensive variability between different species and higher taxonomic categories. Chromosome numbers are known for 1,047 Neotropical freshwater species and 109 marine species, ranging from 2n = 20 in *Pterolebias longipinnis* to 2n = 134 in *Corydoras aeneus* (Oliveira et al., 2007). In general, most Neotropical fish families contain species for which some karyotypic data are available, which shows different evolutionary trends and patterns. When linked to biological features, evolutionary time and geomorphological history, the karyotypic variability can be better understood, allowing for inferences about their evolution and diversity (Oliveira et al., 2007).

In this context, some fish groups have certainly been used as models and are deeper studied. One of the more didactic models that illustrate the importance of cytogenetics in identifying cryptic components of biodiversity is undoubtedly the family Erythrinidae. In addition to conventional chromosome studies, molecular cytogenetic analyses were used to find new chromosomal characteristics for comparative genomics and to provide insights into karyotype differentiation inside the "species complex" of *H. malabaricus* and *E. erythrinus*, which are widespread in the continental waters of South America.

4. Erythrinidae - A fish family as an example for investigating biodiversity

4.1 The Erythrinidae family: General features

The characiform fish family Erythrinidae is a small group composed of three recognized genera, *Hoplias*, *Hoplerythrinus* and *Erythrinus*. It is widespread throughout South America, with a remarkable preference for a great variety of lentic environments such as small and large rivers and lagoons (Oyakawa, 2003) **(Figure 3)**. They are typically carnivorous fish, and several species are broadly distributed throughout the main South America hydrographic basins. The family Erythrinidae has likely a close relationship to the families Lebiasinidae, Ctenoluciidae and the African Hepsetidae (Vari, 1995). Due to their sedentary habits, they are not able to overcome obstacles such as waterfalls and large rapids, which apparently contribute to a reduced gene flow between populations in the same hydrographic river basin.

The fish diversity found in the Neotropics hinders the real definition of many species. In fact, several species have the karyotype described, but have been identified only until the genus level (Oliveira et al., 2007). The taxonomy of the Erythrinidae fishes also reflects this trouble and is still to be better resolved. All the genera appear to have a number of not described species, nowadays included in a same nominal species. Despite the revision for some species of the *Hoplias lacerdae* group (Oyakawa & Mattox, 2009), no revision studies is available for the remaining *Hoplias* species, as well as for the *Hoplerythrinus*, and *Erytrinus* genera.

The erythrinids are fishes that, in general, possess large karyotypic variation (Bertollo et al., 2000; Giuliano-Caetano et al., 2001; Diniz & Bertollo, 2003) and represent excellent models for exploring biodiversity through cytogenetic investigations and for understanding the mechanisms of genomic diversity. The initial cytogenetic studies in this group, mainly based on Giemsa-stained chromosomes, showed the presence of intra-specific variations, with extensive karyotype diversity found among populations in terms of the diploid chromosome number (2n), karyotype composition and different sex chromosome systems. In addition to conventional chromosome studies, molecular cytogenetic analyses proved useful in identifying new cytogenetic characteristics for comparative genomics **(Table 1)**.

4.2 Chromosomal and karyotype diversification among Erythrinidae fishes

The genus *Hoplerythrinus* contains three species, *H. cinereus, H. gronovii* and *H. unitaeniatus* (Oyakawa, 2003), however *H. unitaeniatus* has been cytogenetically analyzed only. A comparative cytogenetic analysis of populations from different Brazilian river basins showed karyotype diversity in this species. Both chromosome number and other karyotypic variations were found among populations, with 2n ranging from 2n = 48 to 2n = 52 and with variable numbers of acrocentric chromosomes (Giuliano-Caetano et al., 2001; Diniz & Bertollo, 2003). However, to date, no heteromorphic sex chromosomes was detected in this species. The available cytogenetic data suggest that *H. unitaeniatus* might include several distinct species and that these fishes require detailed taxonomic analysis to reveal their actual systematic diversity.

The genus *Hoplias* is composed of two large "species groups" (*H. malabaricus* and *H. lacerdae*). The *H. lacerdae* group includes six recently recognized species: *H. brasiliensis, H. aimara,*

Fig. 3. Distribution of *Hoplias malabaricus* karyomorphs A-G (circles); *Hoplias lacerdae* species group (stars); *Hoplerythrinus unitaeniatus* karyomorphs A-D (triangles) and *Erythrinus erythrinus* karyomorphs A-D (squares) in the South America. The large open circles indicate some of the sympatric conditions already detected among distinct *H. malabaricus* karyomorphs.

Species/ Diploid number	Karyotype	Sex chromosomes
Hoplias malabaricus		
A 2n=42	♀♂ 42 m/sm	Not differentiated
B 2n=42	♀ 40 m/sm + 2 st	XX
	♂ 41 m/sm + 1 st	XY
C 2n=40	♀ 40 m/sm	XX
	♂ 40 m/sm	XY
D 2n=40/39	♀ 40 m/sm	$X_1X_1X_2X_2$
	♂ 39 m/sm	X_1X_2Y
E 2n=42	♀♂ 40 m/sm + 2a	Not differentiated
F 2n=40	♀♂ 40 m/sm	Not differentiated
G 2n=40/41	♀ 40 m/sm	XX
	♂ 40 m/sm	XY_1Y_2
Erythrinus erythrinus		
A 2n=54	♀♂ 46 a + 2 st + 6 m	Not differentiated
B 2n=54/53	♀ 46 a + 2 st + 6m	$X_1X_1X_2X_2$
	♂ 44 a + 2 st + 7m	X_1X_2Y
C 2n=52/51	♀ 38 a + 6 st + 8 m/sm	$X_1X_1X_2X_2$
	♂ 36 a + 6 st + 9 m/sm	X_1X_2Y
D 2n=52/51	♀ 44 a + 2 st + 6 m/sm	$X_1X_1X_2X_2$
	♂ 42 a + 2 st + 7 m/sm	X_1X_2Y
Hoplerythrinus uniteniatus		
A 2n=48	♀♂ 48 m/sm	Not differentiated
B 2n=48	♀♂ 46 m/sm + 2 a	Not differentiated
C 2n=52	♀♂ 46 m/sm + 6 a	Not differentiated
D 2n=52	♀♂ 44 m/sm + 2 st + 4 a	Not differentiated
Hoplias lacerdae group		
2n=50	♀♂ 50 m/sm	Not differentiated

Table 1. Karyotype data for the Erythrinidae fish family. m = metacentric; sm = submetacentric and a = acrocentric chromosomes

H. currupira, H. intermedius, H. australis and *H. lacerdae* (Oyakawa & Mattox, 2009). This group of species has a conserved karyotype with an invariable diploid chromosome number (2n = 50) and a karyotype with m and sm chromosomes but no morphologically differentiated sex chromosomes (Bertollo et al., 1978; Morelli et al., 2007; Blanco et al., 2011). Therefore, it seems that speciation in the *lacerdae* group was not accompanied by observable changes at the chromosome level. Such conserved karyotypes and other chromosomal characteristics represent an exception for Erythrinidae fish as compared with the huge karyotypic diversity that has generally been observed in other species of this family.

H. malabricus and *E. erythrinus* are the cytogenetically most studied fish in the Erythrinidae family. Both species have a large geographic distribution in different river basins isolated by millions of years. Conventional and molecular chromosomal markers have proved to be useful indicators for identifying cryptic species diversity. Previous extensive cytogenetic comparative studies showed that many local populations have different karyotypes and other chromosomal characteristics, with a broad range of sex chromosome systems. The overall cytogenetic data clearly suggest that these are an assemblage of species with unresolved taxonomy (Bertollo, 2007) placed under one name that evidently represents a catch-all taxon. More recently, the chromosomal mapping of repetitive DNA families combined with chromosomal painting analysis has improved the understanding of the evolutionary mechanisms involved in the generation of the complex genomic variability in these fish.

4.2.1 *Erythrinus erythrinus*

The genus *Erythrinus* contains only two species, *E. kesslei* and *E. erythrinus* (Oyakawa, 2003), and cytogenetic analyses are available for the later only.

Conventional cytogenetic analyses revealed the presence of extensive karyotype diversity among and within populations of the four currently identified karyomorphs (A to D) (Bertollo et al., 2004). Karyomorph A is composed of populations with 2n = 54 , which have very similar karyotypes composed of 6 metacentric (m), 2 subtelocentric (st) and 46 acrocentric (a) chromosomes and have an absence of differentiated sex chromosomes. Karyomorphs B, C and D share an $X_1X_1X_2X_2/X_1X_2Y$ sex chromosome system, but they differ in their number of chromosomes and karyotype composition. While karyomorph B has 2n = 54 (6m + 2st + 46a) chromosomes in females and 2n = 53 (7m + 2st + 44a) in males, both karyomorphs C and D have 2n = 52/51 but differ in their karyotypes, with 6m + 2sm + 6st + 38a in females and 7m + 2sm + 6st + 36a in males of karyomorph C and 4m + 2sm + 2st + 44a in females and 5m + 2sm + 2st + 42a in males of karyomorph D **(Figure 4)**. The prevalence of acrocentric chromosomes is a particular characteristic that differentiates karyotypes of *E. erythrinus* from those of other erythrinids (Bertollo et al., 2000; Giuliano-Caetano et al., 2001).

The most frequent chromosome number for characiform fish is 2n = 54, and this number likely represents their ancestral diploid chromosome number (Oliveira et al., 2007). Karyomorph A of *E. erythrinus*, with a diploid chromosome number of 2n = 54 and a karyotype dominated by acrocentric chromosomes with undifferentiated sex chromosomes, may represent the most ancestral karyotype among representatives of Erythrinidae. In this view, the occurrence of a lower chromosome number, an increase in the proportion of biarmed chromosomes in the karyotype and the presence of differentiated sex chromosomes represent derived characteristics in the members of karyomorphs B–D. It was hypothesized that karyomorphs B–D resulted from centric fusions between two non-homologous acrocentric pairs producing the submetacentric chromosomes found in their karyotypes and from pericentric inversions, which together generated the observed karyotypic differentiation (Bertollo et al., 2004).

Additional comparative studies of karyomorphs A and D using cytogenetic mapping of repetitive DNA, such as rDNA repeats, satellite DNA, telomeric sequences and classes of TEs, demonstrated that chromosomal rearrangements and genomic modifications were significant events during the course of karyotypic differentiation of this fish. The presence of Interstitial Telomeric Sequences (ITS) in the centromeric region of the only submetacentric chromosome pair found in karyomorph D indicated that a centric fusion created this pair, which is not found in karyomorph A (Cioffi et al., 2010). The most remarkable difference

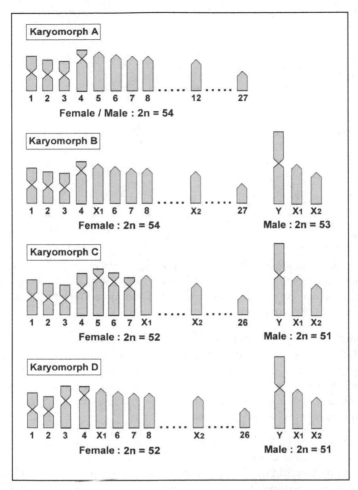

Fig. 4. Extensive chromosomal variability found in the fish *Erythrinus erythrinus*. Partial idiograms of karyomorphs A-D showing their well-defined differences regarding the diploid number, chromosomal morphology and sex chromosome systems.

between karyomorphs A and D was the distribution of 5S rDNA/*Rex3* sites. These sequences co-localized to the centromeric region of several chromosomes. However, while a single chromosome pair was found to bear these sites in karyomorph A, a surprisingly large number of these sequences were found in karyomorph D, with 22 sites in females and 21 in males. Thus, a huge dispersal of 5S rDNA/*Rex3* elements throughout the centromeric regions of the acrocentric chromosomes had occurred; the retroelement *Rex3* might have inserted into a 5S rDNA sequence, giving rise to a 5S rDNA-*Rex3* complex that then moved and dispersed the complex in the genome (Cioffi et al., 2010). Taking into account that karyomorph D represents a derived form as compared to karyomorph A, the chromosomes of this karyomorph may have undergone further rearrangements during the evolutionary process mediated by retrotransposon activity **(Figure 5).**

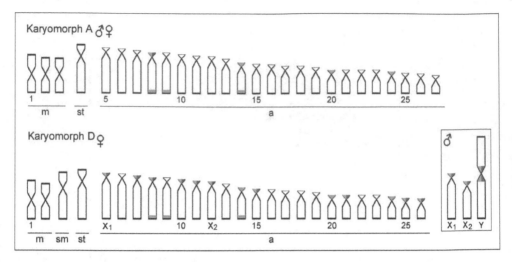

Fig. 5. Representative idiogram of *Erythrinus erythrinus* karyomorphs A and D highlighting the chromosomal distribution of the 18S rDNA (blue) and 5S rDNA/*Rex3* sequences (green/red) in karyomorph A and the expanded distribution of the latter sites in karyomorph D. The sex chromosomes are boxed.

Another remarkable characteristic of *E. erythrinus* karyotypes is the multiple $X_1X_1X_2X_2/X_1X_2Y$ sex chromosome system shared by karyomorphs B–D. Bertollo et al. (2004) proposed that a centric fusion between two non-homologous acrocentric chromosomes might have created the large metacentric Y chromosome and, consequently, the unpaired X_1 and X_2 chromosomes in the male karyotypes, as this sex system appears to have originated before the divergence of these three karyomorphs.

A comparative analysis of male and female karyotypes clearly indicated that the large metacentric Y chromosome originated by a centric fusion harboring characteristic ITS in its centromeric region. Accordingly, the resulting non-homologous acrocentric chromosomes in the male karyotype correspond to the X_1 and X_2 chromosomes (Cioffi et al., 2011a). Chromosome painting also suggested that the X_1X_2Y sex system of *E. erythrinus* was derived from an XY sex pair still morphologically undifferentiated, as found in karyomorph A (Cioffi et al., 2011a), for which there seems to be no apparent specific markers (Cioffi et al., 2011b).

4.2.2 *Hoplias malabaricus*

H. malabaricus cytogenetic analyses showed that several populations possess different karyotypes and other chromosomal characteristics (Bertollo et al., 2000). This species is well adapted to life in small populations with low vagility, which may facilitate the stochastic fixation of chromosomal rearrangements (Faria & Navarro, 2010). Currently, seven karyomorphs (A to G) were easily identified by their number of chromosomes, karyotypes and the presence or absence as well as the size of heteromorphic sex chromosomes (Bertollo et al., 2000).

Conventional cytogenetic analyses were able to distinguish two major karyotype groups in *H. malabaricus*, one consisting of karyomorphs A, B, C, and D (Group 1) and the other containing karyomorphs E, F, and G (Group 2) **(Figure 6)**. Despite their differences in chromosome number, karyomorphs A-D have fairly similar karyotypes, which are different from those of karyomorphs E-G (Bertollo et al., 2000). To date, karyomorphs A-D are the best analyzed. Biogeographical data clearly showed that, while karyomorphs A and C have a wide distribution, karyomorphs B and D are endemic to particular regions (Bertollo et al., 2000; Cioffi et al., 2009). It is noteworthy that some karyomorphs, mainly those ones showing a wider geographical distribution, can be found in sympatric or even in syntopic situations in Brazilian and some other South American regions **(Figure 3)**. In all these cases, no apparent hybrid forms were found suggesting the reproductive isolation between the karyomorphs and, in this way, reinforcing the occurrence of a species complex (Bertollo et al., 2000). Additional RAPD-PCR genomic markers also demonstrated the lack of genetic flow between karyomorph pairs A-C and A-D (Degam et al., 1990), which is compatible with the karyotypic data.

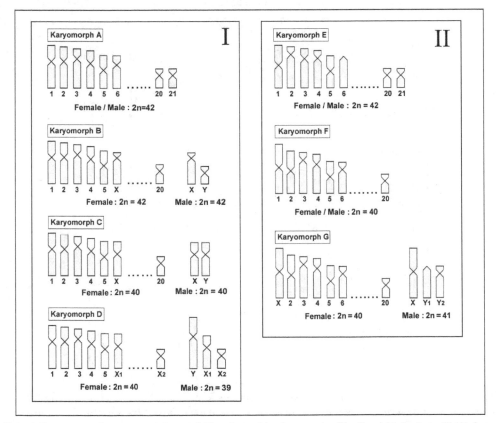

Fig. 6. Extensive chromosomal variability found in the species *Hoplias malabaricus*. Partial idiograms of the karyomorphs A-D (Group I) and E-G (Group II), showing their well-defined differences regarding diploid number, chromosomal morphology and sex chromosome systems.

The *in situ* investigation of repetitive DNA sequences added new informative characteristics that are useful for comparative genomics at the chromosomal level, providing insights into the cytogenetic relationships among *H. malabaricus* karyomorphs A-D. The cytogenetic mapping of repetitive DNA classes (rDNA repeats, satellite DNA, telomeric sequences, several TEs and microsatellite repeats) has provided useful chromosomal markers, highlighting the close relationship among these four karyomorphs and giving additional support to the proposition that they constitute a closely related evolutionary group within *H. malabaricus* **(Figure 7).** The use of repetitive DNA sequences as probes for FISH analyses

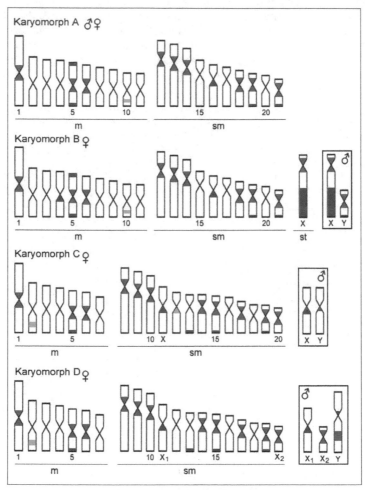

Fig. 7. Representative idiogram of *Hoplias malabaricus* karyomorphs A-D, highlighting the distribution of different classes of repetitive DNA. The locations of the satellite 5S *Hind*III-DNA, 18S rDNA and 5S rDNA sites on the chromosomes are indicated in red, blue and green, respectively. Note that several chromosomes bearing these repetitive DNA sequences were shared by all karyomorphs alongside some karyomorph-specific chromosomal markers. The sex chromosomes are boxed.

has greatly contributed to the study of fish karyotypic evolution, not only because this method provides additional information about the structure of their chromosomes but also because it allows the comparison of genomes of different species (Nanda et al., 2000). In fact, the repetitive DNA fraction of the genome was effective in identifying significant genomic changes that occurred during the differentiation of these karyomorphs of H. malabaricus.

The presence of different sex chromosome systems is also a significant characteristic of the H. malabaricus genome. Three karyomorphs (A, E and F) lack identifiable heteromorphic sex chromosomes, whereas three others (B, D and G) possess well-differentiated sex chromosome systems, an XX/XY system in karyomorph B, an $X_1X_1X_2X_2/X_1X_2Y$ system in karyomorph D and an XX/XY_1Y_2 system in karyomorph G (Bertollo et al., 2000), in addition to an early differentiated XX/XY sex chromosome system found in karyomorph C (Cioffi & Bertollo, 2010). Variation in the amount of several types of repetitive DNA has been shown to be associated with sex chromosome evolution in H. malabaricus. Remarkably, a clear tendency of sex chromosomes to accumulate repetitive DNA was demonstrated (Cioffi et al., 2009, 2010; Rosa et al., 2009). In general, karyomorphs that possesses well-differentiated sex chromosomes (B, D and G) show a restricted geographical distribution, indicating their derived origin. Chromosome painting using whole sex chromosomes as probes has helped to determine the origin of the sex chromosomes in H. malabaricus. Homology was demonstrated between specific chromosomes of karyomorphs A and B as well as between specific chromosomes of karyomorphs C and D (Cioffi et al., 2011c), indicating that the sex systems evolved independently in the different karyomorphs of H. malabaricus. Undoubtedly, this is an important feature considering that the presence of distinct sex chromosome systems might represent a determining factor for the reproductive incompatibility between karyomorphs.

5. Conclusion

Cytogenetics as a whole, as well as fish cytogenetics, has experienced major methodological advances over the years. Much progress has been made in the chromosomal analysis of fish in general, and particularly from the pioneering studies of Neotropical fishes in the 70's. In fact, several improvements in cytogenetic methodologies, specially the advances in molecular cytogenetics, have added to our understanding of chromosomal evolution. Particularly, the mapping of specific DNA sequences on chromosomes by FISH and chromosome painting have proved to be powerful tools. The use of such methodologies has enhanced studies on the molecular composition of the chromosomes and the mechanisms that led to the significant karyotypic differentiation observed in fish. In this context, the Neotropical Erythrinidae family was chosen to illustrate how classical and molecular cytogenetic techniques were useful for comparing the degree of chromosomal diversity over the geographical range of a species, providing important tools for evolutionary and taxonomic studies and increasing knowledge of genomic diversification. The data obtained to date with the use of classical cytogenetic methods and the additional improvements provided by molecular cytogenetics highlighted the hidden biodiversity in distinct species of the Erythrinidae family. Thus, in at least three nominal species of this family, i.e., Hoplias malabaricus, Erythrinus erythrinus and Hopletythrinus unitaeniatus, the karyotype and chromosomal diversity points to the existence of a set of distinct species under evolutionary species concept rather than single biological entities, that are widely distributed throughout the Neotropical region. However, the cases highlighted in the erythrinid fishes can also be found in many other fish species. Chromosomal studies with fishes from different regions of the world have provided reliable information on the inherent diversity of this group. Thus, cytogenetics revealed a powerful tool for discovering biodiversity, with useful applications in evolutionary, taxonomic, phylogenetic and conservation studies.

6. Acknowledgements

Research supported by the Brazilian agencies FAPESP (Fundação de Amparo à Pesquisa do Estado de São Paulo), CNPq (Conselho Nacional de Desenvolvimento Científico e Tecnológico) and CAPES (Coordenação de Aperfeiçoamento de Pessoal de Nível Superior). The authors thank the valuable considerations and suggestions made by Dr. Petr Ráb (Institute of Animal Physiology and Genetics, Libèchov, Czech Republic) in the original manuscript.

7. References

Albert, J. S. & Reis, R. E. (2011). Historical Biogeography of Neotropical Freshwater Fishes, University of California Press, ISBN 978-0520268685, Berkeley, USA

Badaeva, E.D., Dedkova, O.S., Gay, G., Pukhalskyi, V.A., Zelenin, A.V., Bernard, S. & Bernard, M. (2007). Chromosomal rearrangements in wheat: their types and distribution. Genome, Vol.50, No.10, (October 2007), pp. 907-926, ISSN 0831-2796

Bertollo, L.A.C., Takahashi, C.S. & Moreira-Filho, O. (1978). Cytotaxonomic considerations on Hoplias lacerdae (Pisces, Erythrinidae). Brazilian Journal of Genetics, Vol.1, No.2, (August 1978), pp. 103-120, ISSN 0100-8455

Bertollo, L.A.C., Born, G.G., Dergam, J.A., Fenocchio. A.S. & Moreira-Filho, O. (2000). A biodiversity approach in the Neotropical Erythrinidae fish, Hoplias malabaricus. Karyotypic survey, geographic distribution of cytotypes and cytotaxonomic considerations. Chromosome Research, Vol.8, No.7, (October 2000), pp. 603-613, ISSN 0967-3849

Bertollo, L.A.C., Oliveira, C., Molina, W.F., Margarido, V.P., Fontes, M.S., Pastori, M.S., Falcão, J.N. & Fenocchio, A.S. (2004). Chromosome evolution in the erythrinid fish, Erythrinus erythrinus (Teleostei: Characiformes). Heredity, Vol.93, No.2, (August 2004), pp. 228-223, ISSN 0018-067X

Bertollo, L.A.C. (2007). Chromosome evolution in the Neotropical Erythrinidae fish family: an overview, In: Fish Cytogenetics, E. Pisano, C. Ozouf-Costaz, F. Foresti, B.G. Kapoor, (Eds.), 195-211, Enfield, ISBN 978-1-57808-330-5, NH, USA

Biémont C. & Vieira, C. (2006). Genetics: junk DNA as an evolutionary force. Nature, Vol.443, (October 2006), pp. 521-524, ISSN 0028-0836

Blanco, D.R., Lui, R.L., Vicari, M.R., Bertollo, L.A.C. & Moreira-Filho, O. (2011). Comparative cytogenetics of giant trahiras Hoplias aimara and H. intermedius (Characiformes, Erythrinidae): chomosomal characteristics of minor and major ribosomal DNA and cross-species repetititve centromeric sequences mapping differ among morphologically indentical karyotypes. Cytogenetic and Genome Research, Vol.132, No. 1-2, (November 2010), pp. 71-78, ISSN 1424-8581

Bonnaccorsi, S. & Lohe, A. (1991). Fine mapping of satellite DNA sequences along the Y chromosome of Drosophila melanogaster: relationships between satellite sequences and fertility factors. Genetics, Vol.129, No.1, (September 1991), pp. 177-189, ISSN 0016-6731

Charlesworth, B., Sniegowski, P. & Stephan, W. (1994). The evolutionary dynamics of repetitive DNA in eukaryotes. Nature, Vol.371, (September 1994), pp. 215-220, ISSN 0028-0836

Cioffi, M.B., Martins, C. & Bertollo, L.A.C. (2009). Comparative chromosome mapping of repetitive sequences. Implications for genomic evolution in the fish, *Hoplias malabaricus*. BMC Genetics, Vol.10, No.34, (July 2009), pp. 1-8, ISSN 1471-2156

Cioffi, M.B. & Bertollo, L.A.C. (2010). Initial steps in XY chromosome differentiation in *Hoplias malabaricus* and the origin of an X1X2Y sex chromosome system in this fish group. Heredity, Vol.105, No.6, (June 2010), pp. 554-561, ISSN 0018-067X

Cioffi, M.B., Martins, C. & Bertollo, L.A.C. (2010). Chromosome spreading of associated transposable elements and ribosomal DNA in the fish *Erythrinus erythrinus*. Implications for genome change and karyoevolution in fish. BMC evolutionary Biology, Vol. 10, No.271, (September 2010), pp. 1-9, ISSN 1471-2148

Cioffi, M.B., Sánchez, A., Marchal, J.A., Kosyakova, N., Liehr, T., Trifonov, V. & Bertollo, L.A.C. (2011a). Cross-species chromosome painting tracks the independent origin of multiple sex chromosomes in two cofamiliar Erythrinidae fishes. BMC Evolutionary Biology, Vol. 11, No.186, (June 2011), pp. 1-7, ISSN 1471-2148

Cioffi, M.B., Molina, W.F., Moreira-Filho, O. & Bertollo, L.A.C. (2011b). Chromosomal distribution of repetitive DNA sequences highlights the independent differentiation of multiple sex chromosomes in two closely related fish species. Cytogenetic and Genome Research, Vol.134, No.4, (August 2011), pp. 295-302, ISSN 1424-8581

Cioffi, M.B., Sánchez, A., Marchal, J.A., Kosyakova, N., Liehr, T., Trifonov, V. & Bertollo, L.A.C. (2011c). Whole chromosome painting reveals independent origin of sex chromosomes in closely related forms of a fish species. Genetica, Vol.139, No.8, (September 2011), pp. 1065-1072, ISSN 0016-6707

Dergam, J.A. & Bertollo, L.A.C. (1990). Karyotypic diversification in *Hoplias malabaricus* (Osteichthyes, Erythrinidae) of São Francisco and Alto Paraná basins. Brazilian Journal of Genetics, Vol.13, No.4, (February 1990), pp. 755-766, ISSN 0100-8455

Devlin, R. H. & Nagahama, Y. (2002). Sex determination and sex differentiation in fish: an overview of genetic, physiological, and environmental influences, Aquaculture, Vol.208, No.3-4, (June 2002), pp. 191-364, ISSN 0044-8486

Diniz, D. & Bertollo, L.A.C. (2003). Karyotypic studies on *Hoplerythrinus unitaeniatus* (Pisces, Erythrinidae) populations. A biodiversity analysis. Caryologia, Vol.56, No.3, (February 2003), pp. 303-313, ISSN 0008-7114

Doolittle, W. F. & Sapienza, C. (1980). Selfish genes, the phenotype paradigm and genome evolution. Nature, Vol.284, (April 1980), pp. 601-603, ISSN 0028-0836

Faria, R. & Navarro, A. (2010). Chromosomal speciation revisited: rearranging theory with pieces of evidence. Trends in Ecology and Evolution, Vol.25,No.11, (Octuber 2010), pp. 660-669, ISSN 0169-5347

Ferguson-Smith, M. A. & Trifonov, V. (2007). Mammalian karyotype evolution. Nature, Vol.8, (December 2007), pp. 950-962, ISSN 0028-0836

Gall, J. G. & Pardue, M. L. (1969). Formation and detection of RNA-DNA hybrid molecules in cytological preparations. Proceedings of the National Academy of Sciences, Vol.63, No.2, (March 1969), pp. 378-383, ISSN 1091-6490

Giuliano-Caetano, L., Jorge, L.C., Moreira-Filho, O. & Bertollo, L.A.C. (2001). Comparative cytogenetic studies on *Hoplerythrinus unitaeniatus* populations (Pisces, Erythrinidae). Cytologia, Vol.66,No.1, (November 2001), pp. 39-43, ISSN 0011-4545

Griffin, D. K., Robertson, L. B. W., Tempest, H. G. & Skinner, B. M. (2007). The evolution of the avian genome as revealed by comparative molecular cytogenetics. Cytogenetic and Genome Research, Vol.117, No.1-4, (July 2007), pp. 64-77, ISSN 1424-8581

Hoffmann, A.A. & Rieseberg, L.H. (2008). Revisiting the impact of inversions in evolution: From population genetic markers to drivers of adaptive shifts and speciation? Annual Review of Ecology and Systematics, Vol.39, (December 2008), pp. 21-42, ISSN 0066-4162

Jeffreys, A.J., Wilson, V. & Thein, S.L. (1985). Hypervariable 'minisatellite' regions in human DNA. Nature, Vol.314, (March 1985), pp. 67-73, ISSN 0028-0836

Jurka, J., Kapitonov, V.V., Pavlicek, A., Klonowski, P., Kohany, O. & Walichiewicz, J. (2005). Repbase update, a database of eukaryotic repetitive elements. Cytogenetic and Genome Research, Vol.110, No.1-4, (July 2008), pp. 462-467, ISSN 1424-8581

Lévêque, C., Oberdorff, T., Paugy, D., Stiassny, M.L.J. & Tedesco, P.A. (2007). Global diversity of fish (Pisces) in freshwater. Hydrobiologia, Vol.595, No.1, (January 2008), pp. 554-567, ISSN 0018-8158

Mayden, R.L., Tang, K.L., Conway, K.W., Freyhof, J., Chamberlain, S., Haskins, M., Schneider, L., Sudkamp, M., Wood R.M., Agnew, M., Bufalino, A., Sulaiman, Z., Miya, M., Saitoh, K. & He, S. (2007). Phylogenetic relationships of *Danio* within the order Cypriniformes: a framework for comparative and evolutionary studies of a model species. Journal of Experimental Zoology Part B: Molecular and Developmental Evolution, Vol.308, No.5, (September 2007), pp. 642-654, ISSN 1552-5007

Mayr, E. (1942). Systematics and the origin of species from the viewpoint of a zoologist, Columbia University Press, ISBN 0674862503, New York, USA.

Myers, N. (1988). Threatened biotas: 'Hotspots' in tropical forests. The Environmentalist, Vol.8, No.3, (October 1988), pp. 187-208, ISSN 1573-2991

Molina, W.F. (2007). Chromosomal changes and stasis in marine fish groups, In: Fish Cytogenetics, E. Pisano, C. Ozouf-Costaz, F. Foresti, B.G. Kapoor, (Eds.), 69-110, Enfield, ISBN 978-1-57808-330-5, NH, USA

Mooi, R.D. & Gill, A.C. (2010). Phylogenies without Synapomorphies –A Crisis in Fish Systematics: Time to Show Some Character. Zootaxa, Vol.2450 (May 2010), pp. 26-40, ISSN 1175-5334

Morelli, S., Vicari, M.R. & Bertollo, L.A.C. (2007). Evolutionary cytogenetics in species of the *Hoplias lacerdae* , Miranda Ribeiro, 1908 group. A particular pathway concerning the other Erythrinidae fish. Brazilian Journal of Biology, Vol.67, No.4, (December 2007), pp. 897-903, ISSN 1519-6984

Motta-Neto, C.C., Cioffi, M.B., Bertollo, L.A.C. & Molina, W.F. (2011). Extensive chromosomal homologies and evidence of karyotypic stasis in Atlantic grunts of the genus *Haemulon* (Perciformes). Journal of Experimental Marine Biology and Ecology, Vol.401, No.1-2, (May 2011), pp. 75-79, ISSN 0022-0981

Nanda, I., Volff, J.N., Weis, S., Körting, C., Froschauer, A., Schmid, M. & Schartl, M. (2000). Amplification of a long terminal repeat-like element on the Y chromosome of the platyfish, *Xiphophorus maculates*. Chromosoma, Vol.109, No.3, (June 2000), pp. 173-180, ISSN 0009-5915

Nelson, J.S. (1999). The species concept in fish biology. Reviews in Fish Biology and Fisheries, Vol.9, No.4, (December 1999), pp. 1-386, ISSN: 0960-3166

Nelson, J.S. (2006). Fishes of the World (4th edition), Inc. Hoboken, ISBN 0-471-25031-7, New Jersey, USA

Oliveira, C., Almeida-Toledo, L.F., Foresti, F. (2007). Karyotypic evolution in Neotropical fishes. In: Fish Cytogenetics, E. Pisano, C. Ozouf-Costaz, F. Foresti, B.G. Kapoor, (Eds.), 111-164, Enfield, ISBN 978-1-57808-330-5, NH, USA

Oyakawa, O.T. (2003). Family Erythrinidae. In: Check List of the Freshwater Fishes of South and Central America, R. Reis, S. Kullander, C. (Eds.), 238-240, EDIPUCRS , ISBN 8574303615, Porto Alegre, Brazil

Oyakawa, O.T. & Mattox, G.M.T. (2009). Revision of the Neotropical trahiras of the *Hoplias lacerdae* species-group (Ostariophysi: Characiformes: Erythrinidae) with descriptions of two new species. Neotropical Ichthyology, Vol.7,No.2, (June 2009), pp. 117-140, ISSN 1679-6225

Pinkel, D., Straume, T. & Gray, J. (1986). Cytogenetic analysis using quantitative, high sensitivity, fluorescence hybridization. Proceedings of the National Academy of Sciences, Vol.83,No.9, (May 1986), pp.2934-2928, ISSN 1091-6490

Pokorná, M., Giovannotti, M., Kratochvíl, L., Kasai, F., Trifonov, V.A., O'Brien, P.C.M., Caputo, V., Olmo, E., Ferguson-Smith, M.A. & Rens, W. (2011). Strong conservation of the bird Z chromosome in reptilian genomes is revealed by comparative painting despite 275 million years divergence. Chromosoma, Vol.120, No.5, (October 2011), pp. 455-468, ISSN 0009-5915

Ráb, P., Bohlen, J., Rábová, M., Flajšhans, M. & Kalous, L. (2007): Cytogenetics as a tool in fish conservation: the present situation in Europe. In: Fish Cytogenetics, E. Pisano, C. Ozouf-Costaz, F. Foresti, B.G. Kapoor, (Eds.), 215-241, Enfield, ISBN 978-1-57808-330-5, NH, USA

Reis, R.E., Kullander, S.O. & Ferraris Jr, C.J. (2003). Check List of the Freshwater Fishes of the South and Central America, EDIPUCRS, ISBN: 8574303615, Porto Alegre, Brazil

Rosa, R., Rubert, M., Vanzela, A.L.L., Martins-Santos, I.C. & Giuliano-Caetano, L. (2009). Differentiation of Y chromosome in the $X_1X_1X_2X_2/X_1X_2Y$ sex chromosome system of *Hoplias malabaricus* (Characiformes, Erythrinidae). Cytogenetic and Genome Research, Vol.127, No.1, (December 2009), pp. 54-60, ISSN 1424-8581

Ruffing, R. A., Kocovsky, P. M. & Stauffer, J. R. (2002). An introduction to species concepts and speciation of fishes. Fish and Fisheries, Vol.3, No.3, pp.143–145, ISSN 1467-2979

Schwarzacher, T. (2003). DNA, chromosomes, and in situ hybridization. Genome, Vol.46, No.6, (December 2003), pp. 953-962, ISSN 0831-2796

Teruel, M., Cabrero, J., Montiel, E. E., Acosta, M. J., Sánchez, A. & Camacho, J.P. (2009). Microdissection and chromosome painting of X and B chromosomes in *Locusta migratoria*. Chromosome Research, Vol.17,No.1, (December 2009), pp.11-18, ISSN 0967-3849

Vari, R. P. (1995). The Neotropical fish familiy Ctenoluciidae (Teleostei: Ostariophysi: Characiformes): supra and intrafamilial phylogenetic relationships, with a revisionary study. Smithsonian Contributions to Zoology, Vol.564, No.6, (April 1995), pp.1-97, ISSN 0081-0282

Venkatesh, B. (2003). Evolution and diversity of fish genomes. Current Opinion in Genetics & Development, Vol.13, No.6, (December 2003), pp.588-592, ISSN 0959-437X

Yang, F. & Graphodatsky, A. S. (2009). Animal probes and ZOO-FISH. In: Fluorescence In Situ Hybridization (FISH): Application Guide, T. Liehr (Ed.), 323-346, Springer, ISBN 3540705805, Berlin, Germany

Permissions

The contributors of this book come from diverse backgrounds, making this book a truly international effort. This book will bring forth new frontiers with its revolutionizing research information and detailed analysis of the nascent developments around the world.

We would like to thank Dr. Padma Tirunilai, for lending her expertise to make the book truly unique. She has played a crucial role in the development of this book. Without her invaluable contribution this book wouldn't have been possible. She has made vital efforts to compile up to date information on the varied aspects of this subject to make this book a valuable addition to the collection of many professionals and students.

This book was conceptualized with the vision of imparting up-to-date information and advanced data in this field. To ensure the same, a matchless editorial board was set up. Every individual on the board went through rigorous rounds of assessment to prove their worth. After which they invested a large part of their time researching and compiling the most relevant data for our readers. Conferences and sessions were held from time to time between the editorial board and the contributing authors to present the data in the most comprehensible form. The editorial team has worked tirelessly to provide valuable and valid information to help people across the globe.

Every chapter published in this book has been scrutinized by our experts. Their significance has been extensively debated. The topics covered herein carry significant findings which will fuel the growth of the discipline. They may even be implemented as practical applications or may be referred to as a beginning point for another development. Chapters in this book were first published by InTech; hereby published with permission under the Creative Commons Attribution License or equivalent.

The editorial board has been involved in producing this book since its inception. They have spent rigorous hours researching and exploring the diverse topics which have resulted in the successful publishing of this book. They have passed on their knowledge of decades through this book. To expedite this challenging task, the publisher supported the team at every step. A small team of assistant editors was also appointed to further simplify the editing procedure and attain best results for the readers.

Our editorial team has been hand-picked from every corner of the world. Their multi-ethnicity adds dynamic inputs to the discussions which result in innovative outcomes. These outcomes are then further discussed with the researchers and contributors who give their valuable feedback and opinion regarding the same. The feedback is then collaborated with the researches and they are edited in a comprehensive manner to aid the understanding of the subject.

Apart from the editorial board, the designing team has also invested a significant amount of their time in understanding the subject and creating the most relevant covers. They scrutinized every image to scout for the most suitable representation of the subject and create an appropriate cover for the book.

The publishing team has been involved in this book since its early stages. They were actively engaged in every process, be it collecting the data, connecting with the contributors or procuring relevant information. The team has been an ardent support to the editorial, designing and production team. Their endless efforts to recruit the best for this project, has resulted in the accomplishment of this book. They are a veteran in the field of academics and their pool of knowledge is as vast as their experience in printing. Their expertise and guidance has proved useful at every step. Their uncompromising quality standards have made this book an exceptional effort. Their encouragement from time to time has been an inspiration for everyone.

The publisher and the editorial board hope that this book will prove to be a valuable piece of knowledge for researchers, students, practitioners and scholars across the globe.

List of Contributors

Sandra Milena Rondón Lagos
Doctoral Program in Biomedical Sciences, Universidad Del Rosario, Colombia

Nelson Enrique Rangel Jiménez
Azienda Ospedaliero-Universitaria S. Giovani Battista di Torino, Italy

Anderson Fernandes
University of Mato Grosso State, Department of Biology/Tangará da Serra, Brazil
Federal University of Viçosa, Department of General Biology/Viçosa, Brazil

Diones Krinski
Federal University of Paraná, Department of Zoology/Curitiba, Brazil

Marla Piumbini Rocha
Federal University of Pelotas, Department of Morphology/Pelotas, Brazil

Danuta Januszkiewicz-Lewandowska
Department of Medical Diagnostics, Poznań, Poland
Institute of Human Genetics Polish, Academy of Sciences, Poznań, Poland
Department of Pediatric Hematology, Oncology and Transplantology of, Medical University, Poznań, Poland

Ewa Mały
Department of Medical Diagnostics, Poznań, Poland

Jerzy Nowak
Institute of Human Genetics Polish, Academy of Sciences, Poznań, Poland

Dorota Kwasny, Indumathi Vedarethinam, Pranjul Shah, Maria Dimaki and Winnie E. Svendsen
Technical University of Denmark, Department of Micro- and Nanotechnology, Denmark

Ricardo Barini, Isabela Nelly Machado and Juliana Karina R. Heinrich
Fetal Medicine Program, Cytogenetics Core, Women's Hospital, State University of Campinas, UNICAMP, Campinas, SP, Brazil

Małgorzata Krawczyk-Kuliś
Medical University of Silesia, Poland

Kimio Tanaka
Department of Radiobiology, Institute for Environmental Science, Aomori, Japan

Marcelo de Bello Cioffi and Luiz Antonio Carlos Bertollo
Universidade Federal de São Carlos, Brazil

Wagner Franco Molina
Universidade Federal do Rio Grande do Norte, Brazil

Roberto Ferreira Artoni
Universidade Estadual de Ponta Grossa, Brazil

Printed in the USA
CPSIA information can be obtained
at www.ICGtesting.com
JSHW011335221024
72173JS00003B/161